ESSENTIAL HYDRAULICS

FLUID POWER BASIC

M. WINSTON

Copyright © 2012 M. Winston

All rights reserved.

ISBN: 148410658X
ISBN-13: 978-1484106587

DEDICATION

To Mabel, my wife;

LJ my son.

With Love.

PREFACE

For many years, it seems that the objective in Engineering Education has been to teach courses and topics to Engineering students to become graduates. The emphasis has been on theoretical topics with very minor practical applications. However, in the real world the majority of engineering graduates would end up as practising engineers or technologists.

This book serves as a starter kit for those who are works or projects related. It is also for all the others who will use it as basic guide book.

This oil hydraulics subject will be published in 2 volumes; 'Basic' and 'Intermediate'. Although both books are intended to include as much details as possible, it can only cover a small tip on the iceberg in term of hydraulic fluid power and controls system. Book I attempts to introduce the very fundamental hydraulic components, symbols, applications and interpret basic hydraulic circuit diagrams.

Book II requires readers with some practical knowledge on hydraulic system and more in-depth than Book I. It covers more projects or works related examples and solutions.

Book I is also intended as a bridging course for anyone who wanted to know about hydraulic systems. The contents of this book especially chapter 1, 2. and 3 are carefully planned for beginners.

Nevertheless, these 2 volumes are compressed and compiled by drawing on the many years of practical experience in diverse fields.

CONTENTS

Book I – Essential Hydraulics - Fluid Power (Basic)

Chapter 1 - Basic Principle of Fluid Flow 3

Pascal's Principle 3
Viscosity 6
Continuity Formula 9
Laminar and Turbulent Flow 11
Pressure Loss 13
Flow Characteristic In Restricted Condition 15

Chapter 2 - Basic Hydraulic Equipment 18

Basic Hydraulic Symbols 18
Fluid Power Element Equipment 19
Hydraulic Pumps 20
Vane Pumps   21
  - Fixed Displacement Pump (Single-stage)
  - Double pumps
  - Combination pump (Two-stage)
  - Variable Displacement Pumps
Gear Pumps (External & Internal) 25
Piston Pumps (Axial & Radial) 28
Hand Pumps 29
Pump Performances 30
Type of Pumps comparison 35
Pump Selection Criteria 37
Practical Situation on Site 38

Chapter 3 – Flow Control Valves 41

Control Valves 42
Check Valves (Non return & pilot operated) 44
Pressure Control Valves 47
Pressure Relief Valves 48
Pressure Reducing Valves 50
Unloading Valves 52

Sequencing Valves 53
Counterbalance Valves 54
Brake Valves 56
Directional Control Valves 57
Classification of Valves by Spool Designs 59
Three-way & Four-way Control Valves 65
Solenoid Directional Control Valves 67
Manual Directional Control Valves 75
Servo Control Valves 76
Flow Control Valves (Meter-in & Meter-out) 77
Shuttle Valves 81
Flow Divider valves 83
Stackable Valves System (CETOP/NFPA) 84
On/Off Valves 85
Manifolds 86

Chapter 4 - Actuators 88

Actuator 88
Hydraulic cylinders 89
 - Tie-rod & welded end design
Type of Hydraulic Cylinders 95
Single Acting Cylinder 95
Double Acting Cylinder 95
Telescopic Cylinder 96
Hydraulic Jack 96
Hydraulic Motors 97
 - vane, gear, piston & oscillating types

Chapter 5 – Hydraulic Basic Accessories 102

Oil Filters and Tank Filters 103
Oil heater/Cooler 104
Measuring Instruments 105
Accumulators and applications 105
 - Weight-Loaded
 - Bladder
 - Diaphragm
 - Piston

Hydraulic Oil **111**
Selection Criteria of Hydraulic Fluid **112**
Viscosity of Hydraulic Fluid **113**
Tank size and Tank Accessories **114**
Electric Motors and selection **116**
Steel Piping and Fluid Power Transmission **118**
Pipe Schedules and Codes **119**
Steel Tubing and Flexible Hoses **121**

Chapter 6 - Power System and Controls **124**

Principles of Control **124**
Basic Electricity **127**
Electrical and Mechanical Energy **128**
Unit of Power **133**
Power Supply - **AC & DC** **135**
AC and DC Power Generators **137**
Solar Photovoltaic (PV) Panels **137**
Power/Energy Storage System **139**
 - Battery Cells
Relays **139**
Solenoids **143**
Controls System **145**
Type of Control Devices **148**
Type of Pressure Switches **149**
Pressure Sensor and Transducers **150**
Temperature Switches and Sensors **152**
Thermocouple **153**

Chapter 7 - Basic Hydraulic System **154**

Basic Hydraulic Power System **154**
 - Sizing - step by step example
Sizing Hydraulic Cylinder **157**
Sizing Suitable Pump **161**
Oil Reservoir Tank & Pump Size **163**
Electric Motor sizing **166**
Other HPS off the shelves **173**

Chapter 8 – Design Procedure and Troubleshooting 175

Hydraulic System & Equipment Design Procedures **176**
Hydraulic System Troubleshooting **177**

Chapter 9 - Advance Technology update 184

Introduction **184**
Data and Digital **185**
Industrial Networks **187**
Ethernet and the Information Highway **188**
Data Transfer Rate **189**
Open System and Proprietary System **190**
Wireless and Mobility **190**

Appendices 193 - 209

Appendix I - Important information and charts 193

Table I.1 - Pressure & Liquid Head **193**
Table I.2 - Pressure/Force of Standard Cylinders **194**
I.2 Common unit Conversion Factors 195
I.3 Output forces and speed of cylinder formulae 196
I.4. Useful Hydraulic Formulae 197

Appendix II - Hydraulic Seals and O-rings 199

II.1 Hydraulic Seals and O-rings 199
Table II.1 - O-rings Material 208
II.2 Three phase current and electric motor 209

ACKNOWLEDGMENTS

I wish to thank all my ex-colleagues, for their encouragement to write this book and their invaluable feedbacks that had helped to shape the contents of this book. I also wish to extend my gratitude to Dr. Andrew for his comments and suggestions.

Chapter 1 - FUNDAMENTALS OF OIL-HYDRAULICS

Pascal's Principle
Viscosity
Continuity Formula
Lamina and Turbulent Flow
Pressure Loss
Flow Characteristic In Restricted Condition

Upon completion of this chapter, you will:
1. Understand basic Pascal's Principle
2. Apply Pascal Principle to examine the advantage of forces in fluid system
3. Know about flow continuity in a closed fluid system
4. Recognise a hydraulic system at work of equipment and machineries using Pascal's Principle

Strive not to be a success, but rather to be of value.
Albert Einstein

Pascal's Principle

The pressure in a static fluid has the following properties:

The pressure works vertically to a wall face.
The pressure at any given point is the same in all directions.
The pressure applied to a part to fluid in a closed container is transmitted uniformly to other parts of the fluid.

These properties are known as Pascal's principle. For instance, Figure 1.1 shows two interconnected cylinders of different diameter filled with liquid.

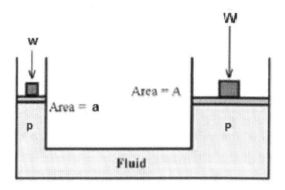

Figure 1.1

When weight w is placed on the smaller cylinder, the pressure $P\ (=\frac{w}{a})$, or weight W divided by sectional area of the cylinder, is produced and transmitted to the larger cylinder.

According to the principle, the pressure increases in proportion to the ratio of sectional area of the larger cylinder to that of the smaller cylinder, or $W = PA = \frac{A}{a}$. This is the most important principle in oil-hydraulics.

Now, we can express **W** as the force **F** applied on the above system in Fluid Mechanics term,

We have,

$$F = PA$$

$$\text{or } P = \frac{F}{A} \quad \ldots\ldots\ldots\ldots \text{(Eq. 1.1)}$$

The pressure is a force applied to unit area. In metric unit, it is in kgf/cm^2. For imperial unit, it is in lb/in^2 (psi). 1 lb/in^2 is approximately equal to 0.07 kg/cm^2, and 1 kgf/cm^2 is to 14.22 lb/in^2 (psi).

Example 1.1

Recall Figure 1.1, if $D_1 = 0.05$ m for Area A_1 and $D_2 = 0.15$ m for A_2, and has a mass of $m_1 = 100$ kg exerted at F_1, then

Using Equation 1.1, $P = \frac{F}{A}$, and knowing fluid Pressure P is the same at any point and direction, that is $P = p$, so we have

$$F_2 = \frac{A_2}{A_1} F_1$$

$$= \frac{\pi(0.15)^2}{\pi(0.05)^2} (981 \text{ N})$$

$$= 8829 \text{ N}$$

$$F_2 = 8829 = (m_2 \text{ kg})(9.81 \text{ m/s}^2)$$

$$= \mathbf{900 \text{ kg } (1{,}984 \text{ lb})}$$

NB: If the load m_1 increases to a value of 200 kg (441 lb), then m_2 will be **1,800 kg (3,969 lb)**

Apparent Weight

Apparent weight problems usually are encountered in the discussion of force. Though relatively easy, they cause problems for some people, the rest may choose to skip this section. This short section should give you enough background and an example to show you how to do apparent weight problems.

One of the first points to get straight is the difference between **mass** and **weight**. *Mass is that property that makes different objects accelerate differently when equal forces are applied.* Operationally mass is the m in the equation **F = ma**. Mass is measured in kilograms (kgs) or slugs.

Weight in equation form is, **W = mg** ..... **(Eq. 1.2)**

and is the force something exerts. Weight is measured in **Newtons or pounds.**

Example 1.2

A person of 80 Kg mass standing on earth is subject to the force of gravity that acts between the 80 kg person and the mass of the earth. This force is expressed as an acceleration due to gravity. On the earth, this acceleration is 9.8 m/s^2 (or 32.2 ft/s^2). So the force on the person, also called weight is

$$F \text{ or } W = 80 \text{ kg} \cdot 9.8 \text{ m/s}^2$$

$$= 784 \text{ N } (176 \text{ lb})$$

Viscosity

When fluids make the relative movements, a force is generated to resist these movements which occurs along a boundary of fluids. This is termed as fluid viscosity.

The Frictional stress **t** working on the boundary is proportional to velocity grade line of an adjacent layer ($\frac{du}{dy}$), as shown in Figure 1.2 and is expressed by the following formula:

$$t = u \frac{du}{dy} \quad \text{....... (Eq. 1.3)}$$

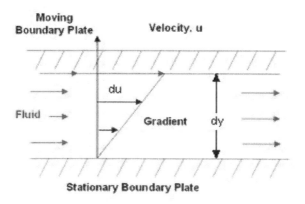

Figure 1.2

Proportional constant **u**, in the formula takes different values according to the type of fluid and it respective temperature, and pressure, and is known as the coefficient of viscosity (u, as viscosity).

The Engineering units of viscosity **u** are $kg.s/m^2$ and c.g.s. units are g/cm.s. 1 g/cm.s is called 1 poise (p), 1/100 of it is 1 centi-poise (c.p).

The engineering units of $kg.s/cm^2$ are converted to c.p units by the following formula:

$$1 \text{ c.p.} = 1.02 \times 10^{-4} \text{ kg.s/m}^2$$

Since a fluid changes in viscosity and density with the change in pressure, the unit of **dynamic viscosity** y is used. This can be determined by dividing viscosity **u** by corresponding density ρ, or simply as follows,

$$y = \frac{u}{\rho}.$$

The unit of dynamic viscosity y, both for engineering and c.g.s unit, is m²/s or cm²/s. 1 cm²/s is called 1 stoke (St), $\frac{1}{100}$ of it is 1 centi-stoke (cSt).

The dynamic viscosity is determined by measuring the time (usually in seconds) required for a gravity flow of a specified amount of fluid by a viscometer.

While cent-stokes are usually used in Japan, Engler degrees are used in France and Germany, Saybolt universal seconds (SUS, or Saybolt seconds universal, SSU) in the USA, and Red wood seconds in the UK.

Approximate conversions of Engler degrees to centi-stokes can be found by the following formula:

$$y = 7.6\ E(1 - \frac{1}{E^2}) c.St \quad \ldots\ldots\ldots\ldots \text{(Eq. 1.4)}$$

Where **E** is Engler degree.

The approximate conversions of Saybolt Universal seconds and

Red wood seconds to centi-stokes can be determined as follows:

$$y = At - \frac{B}{t} \text{ c.St} \quad \text{............... (Eq. 1.5)}$$

Where t is the number of seconds in each viscosity, and A and B are the coefficients in each viscosity. (**A = 0.22, B = 180** for Saybolt Universal seconds, and **A = 0.26, B = 171** for Red wood seconds).

Continuity Formula

When an incompressible fluid flows continuously in a pipe as depicted in Figure 1.3, the following equation is derived with the law of conservation of mass:

$$\mathbf{A_1.v_1 = A_2.v_2 = Q = const.} \quad \text{........... (Eq. 1.6)}$$

Where A_1 and A_2 are the areas of given cross sections 1 and 2, v_1 and v_2 are the mean velocities of flow at the cross sections 1 and 2, and the rate of flow passing the pipe is **Q**.

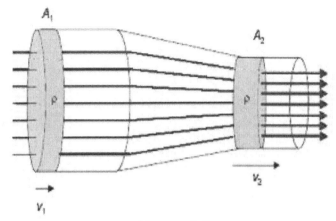

Figure. 1.3

This is known as the continuity equation. The flow rate **Q** is expressed as the product of the cross sectional areas and the mean velocities.

The rate of flow is defined as the quantity of fluid passing within a unit time. In metric unit, it is measured in l/min, also described as the volume flow rate in Fluid Mechanics term. For imperial unit, it is gal/min (gpm). 1 gal/min is equivalent to about 3.785 l/min, and 1 l/min is approximately 0.264 gpm. (NB: US gallon).

Laminar and Turbulent Flow

Flow of oil in a pipe is classified as a closed conduit. It is extremely important to examine the condition of oil flow in pipe for hydraulic system. In any enclosed situation, the fluid flow can be in the form of laminar or turbulent.

When laminar flow exists, the fluid flows smoothly through the pipe in layers called *laminae*. A fluid particle in one layer stays in that layer. When turbulent flow exists, the flowing fluid particles move about the cross section of the pipe. These are the eddies and vortices responsible for mixing action. Such eddies and vortices do not exist in laminar flow.

Figure 1.4 shows the two types of flow in pipe.

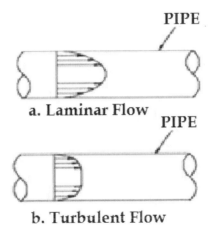

Figure 1.4

For the laminar flow in pipe, it is subjected to loss due to viscosity of oil only, whereas in the turbulent flow it losses can be due to the shape and roughness of pipe wall.

For this reason, it is desirable to maintain laminar flow at all time.

Flow in pipes may be expressed by Reynolds number (Re) which is determined as follows:

$$\mathbf{Re} = \frac{vD}{y} = \frac{4Q}{\pi D y} \quad \text{(Eq. 1.7)}$$

Where **Re** is Reynolds number,

v is the mean velocity (cm/sec),
D is the inner diameter of pipe (cm),
Q is the rate of flow (cm^3/sec), and
y is the coefficient of dynamic viscosity (cm^2/sec.St).

The Reynolds number at which a flow changes from laminar flow to turbulent flow is called the critical Reynolds number, below which laminar flow is maintained above which turbulent flow occurs. The velocity in the critical Reynolds number is called the critical velocity. The critical Reynolds number takes different values depending on the nature of pipe line.

For smooth circular pipes, the value ranges between 2,000 and 2,300. On the other hand, for oil hydraulics circuits using joints, bent pipes and valves, it is safer to think that the critical Reynolds number is around 1,000.

Pressure Loss

When a fluid moves in pipe, friction due to viscosity of fluid occurs between fluid and solid body, acting as the resistance to fluid flow. In addition to the friction resistance, pressure loss occurs at entrances or exits of pipes, at points where cross section of pipes changes, pipes bend, or pipes branch out or joint together, or at valves or orifices. With high pressure system, this results in frictional heating.

In other words, kinetic energy of oil is wasted through pressure loss when flowing through pipes and valves of cylinders.

Therefore, it is absolutely necessary to minimize pressure losses in oil hydraulic circuits in order to maintain circuit efficiency at all times. Too much of pressure losses in the system would result in delivering the correct pressure to the load ends. It not only contribute to unnecessary heating that affects oil temperature, it also affect the viscosity of the oil. Over times, the higher oil temperature would cause premature failure to important seals of hydraulic components.

Pressure loss in steady flow in a smooth and regular circular pipes is generally expressed by Darcy's formula as follows:

$$\Delta P = (\lambda \rho)(\frac{L}{D})(\frac{v^2}{2g}) \text{ kgf/cm}^2 \quad \text{.................. (Eq. 1.8)}$$

Where ΔP is the pressure loss (kgf/cm2),
L is the pipe length (cm),
D is the inner diameter of pipe (cm),
v is the mean velocity (cm/s),
g is the gravity acceleration (980 cm/sec^2).
ρ is the density of oil (kg/cm^3), and
λ is the friction coefficient of pipe.

Friction coefficient of pipe λ is a function of Reynolds number **Re** and surface roughness of pipe walls.

For smooth circular pipes with laminar flow (**Re< 2 x 1000**), λ may be expressed as follows:

$$\lambda = 64/Re \quad \text{............... (Eq. 1.9)}$$

For those with turbulent flow in the range of **2.5 x 1000 < Re < 10^4**, λ may be expressed as follows:

$$\lambda = 0.3164\, Re^{-0.25} \quad \text{............... (Eq. 1.10)}$$

A primary factor to determine the relationship between the diameter of oil pressure control valve and the rated flow is the degree of pressure loss at the valve. In general, flow rate with pressure loss of 3 kgf/cm^2 (42 psi) is fairly common in hydraulic and known as the rated flow.

Flow Characteristic In Restricted Conditions

Figure 1.5 shows 2 types of flow characteristics commonly encounter in an Orifice Restriction.

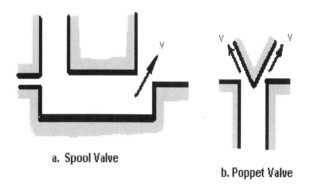

a. Spool Valve

b. Poppet Valve

Figure 1.5

The definition of an orifice restriction refers to the restriction part which is shorter than cross sectional size. Examples of the hydraulic fluid undergoing such restriction parts can be found in spool valves and poppet valves. Flow rate **Q** through this type of restriction is given by the following formula:

$$Q = cA \sqrt{\left(\frac{2g}{\rho}\right) \Delta P} \text{ (cm}^3/\text{sec)} \dots \dots \text{ (Eq. 1.11)}$$

Where **c** is the flow coefficient,

A is an area of orifice (cm²),
ρ is the specific gravity of oil (kg/cm), and
ΔP is the difference in pressures before and after the orifice (**kgf/cm²**).

Flow coefficient **c** is an important indicator of characteristic in restriction parts, and where Reynolds number is large; it takes almost a constant value according to the degree of contraction of low jetting from the orifice.

For orifice restriction, the resistance is less affected by oil temperature as well as by viscosity.

Choke Restriction

Choke restriction is applied to flow restriction in which the length of flow is longer than its cross section, such as a small hole provided in a balance piston of pilot controlled relief valve. Example of such flow is shown in Figure. 1.6.

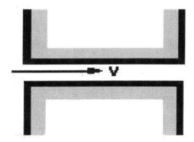

Figure 1.6

Flow rate **Q** in choke restriction for circular cross section is given as follows:

$$Q = \frac{\pi D^4 g \Delta P}{128 \, \rho \, y \, L} \text{ cm}^3/\text{sec} \quad \ldots\ldots\ldots\ldots\ldots \text{(Eq. 1.12)}$$

Where,

D is the diameter of hole (cm),
y is the dynamic coefficient of viscosity of oil (cm²/sec), and
L is the length of hole (cm).

In choke restriction, the resistance is greatly affected by oil temperature and as well as by its viscosity.

Chapter 2 - Basic Hydraulic Equipment

Basic Hydraulic Symbols
Fluid Power Element Equipment
Hydraulic Pumps
Vane Pumps
 - **Fixed Displacement Pump (Single-stage pump)**
 - **Double pumps**
 - **Combination pump (Two-stage pump)**
 - **Variable Displacement Pumps**
Gear Pumps (External & Internal)
Piston Pumps (Axial & Radial)
Hand Pumps
Pump Performances
Type of Pumps comparison
Pump Selection Criteria
Practical Situation on Site

Upon completion of this chapter, you will be able to:
1. Recognise the basic hydraulic symbols
2. The various types of hydraulic pumps
3. Know about each type of pump capabilities, capacities and functions
4. Recognise and identify the type of pumps used on site

For every life, there is a flow. The more you ask, the more you'll know. The more you give, the more you'll grow.
Herman Melville

BASIC HYDRAULIC SYMBOLS

Most hydraulic equipment is made up of several hydraulic components that work together. It would be very tedious and sometimes almost impossible to explain how the hydraulic system and control operates without using circuit diagrams. Therefore, fluid power system must also have a set of universal symbols to represent each hydraulic component, similar to that of electrical and electronics symbols, just that the hydraulic ones are less complicated. Common components are easy to identify. It is also easy to learn the symbols used to represent the hydraulic components.

Here, we will present the very fundamental ones to facilitate your basic understanding. When you have completed with this book, you should be able to design and produce hydraulic circuit diagrams confidently on your own.

With more practices, you will also notice very quickly that all these basic-element symbols and representations replicate in other sophisticated valves' function. Perhaps, if you do come across one, just apply what you have learnt in this book. Chances are you won't be far from wrong. After all, we all learnt new things by taking a bold step.

NB: Further references of other symbols can be obtained with the manufacturers and at ISO 1219. Note also ISO 1219 is not free.

FLUID POWER ELEMENT EQUIPMENT

The basic fluid power system consists of a source of input energy and a suitable device for energy input, energy output, and energy modulation. Transmitting fluid power requires a pump to convert the mechanical energy into fluid energy. The primary source of input energy is quite often an electric motor or an internal combustion engine or other type of mechanical devices that can supply force and motion to operate the pump. The pump supplies hydraulic fluid pressure to the system. In other words, hydraulics is defined as the generation of forces and motion using hydraulic fluids. The hydraulic fluids represent the medium for power transmission.

Hydraulic Pumps

There are 4 major types of pumps in oil-hydraulic system;

1. Vane Pumps
2. Gear Pumps
3. Piston Pumps
4. Hand Pumps

Hydraulic Pumps	Hydraulic Symbols	
	Single Direction Flow	
	Fixed	Variable
Vane Pumps Gear Pumps Piston Pumps (Axial & Radial Types) Hand Pump	⊙⊣	⌀⊣
	Double Direction Flow	
	Fixed	Variable
	⊙⊣	⌀⊣

Figure 2.1

Hydraulic pumps are used to convert mechanical power supplied from external source to fluid power. Most of hydraulic pumps are of **displacement** type. They are classified as vane type, gear type and piston type, and rated according to their service pressure as listed in Table 2.2.

Table 2.1 – Pressure conversion (SI to Imperial)

| \multicolumn{11}{c}{1 kg/cm² = 14.223 psi} |

Kgf/cm²	35	70	105	140	175	210	245	280	315	350
psi	500	1,000	1,500	2,000	2,500	3,000	3,500	4,000	4,500	5,000

1 kg/cm² = 14.223 psi
1 psi = 0.0703 kg/cm²
1 bar = 14.5 psi

VANE PUMPS

Common Types of Vane Pump used in the industries are;

1. Fixed Displacement Pump – Single-stage pump
2. Double pump
3. Combination pump - Two-stage pump
4. Variable Displacement Pump

S/N	Descriptions	Physical Appearance	Sectional View
1.	Vane Pump	DOUBLE SINGLE	
2.	Hydraulic Symbol	OUTLET 1 OUTLET 2 INLET	

Fig 2.2

Advantages of vane pump

1. Less ripple in discharge pressure (practically, the ripple does not need to be taken into account).
2. Efficiency deceases to a lesser extent even if the vane is worn out.
3. Small dimensions relative to the hydraulic power of pump.
4. Ease in maintenance due to less failure.

Single-stage pumps

There is a variety of single-stage pumps available, in terms of pressure and displacement, and they are most widely used among other type of pumps as hydraulic power source.

Low pressure single-stage pumps

Low pressure single-stage pumps have maximum operating pressure range between 50 kgf/cm^2 (711 psi) and 70 kgf/cm^2 (995 psi) and are being used for a variety of purposes, from small machine tools to large steel making machinery.

Example of pump Performance

Speed: 1,200 rpm

Viscosity: 40 cSt (190 SSU)

High pressure single-stage pumps

High pressure single-stage pumps have maximum operating pressure range between 160kgf/cm^2 (2,275 psi) and 210 kgf/cm^2 (2,986 psi) and are widely used for high pressure equipment such as injection moulding machine.

There are of pressure loading type construction to maintain a sufficient efficiency even if the pressure increases. The displacement capacity ranges between 6 to 237 cm^3/rev (0.365 to 14.45 cu. in/rev).

Combination pumps

Combination pumps incorporate 2 units of pumps with other operating element such as, relief valves, unloading pressure control valves and check valves. These components formed an integral functional unit to control each equipment in relation to each other according to load.

In addition, combination pumps employ an unique mechanism to reduce the wear of cam ring and vane due to rubbing.

Double pumps

Double pumps are formed by connecting 2 units of pumps in series operating with the same drive shaft. Each pump can be operated separately, and it can be combined for low pressure and high pressure operations. The combination of any displacement mixing of two pumps can be freely selected. Double pumps series can be made up of vane or gear pumps.

S/N	Descriptions	Physical Appearance
1.	Double Pump	
2.	Hydraulic Symbol	

Figure 2.3

Variable displacement pumps

In variable displacement pumps, the discharge can be reduced, when the pressure reaches a **pre-set level.** It then automatically regulates the rate of flow required in the system and at the same time maintaining the pressure. In this class, pressure compensator type is generally used in most cases, as it can be described as a high power efficiency circuit since its input does not increase proportionally to the pressure unlike the fixed displacement pump.

However, a small limitation on the variable displacement pumps is that it have pressure imbalance. Thus the load on bearing increases in proportion to pressure, while fixed displacement pumps are of pressure balance type and they do not have radial load works on the bearing. This pump commonly use in mobile industrial applications.

The variable displacement pump can be unloaded by either of the following two methods:

1. Pressure at maximum, discharge at zero – Fully close the discharge pipe.

2. Pressure at minimum, discharge at maximum – Fully open the discharge pipe.

Method (2) is preferable in consideration of pump's durability, temperature rise and noise. In addition, sudden surge in pressure on the system can happen during the shift caused by directional control, although not damaging. It needs to be addressed in some cases.

GEAR PUMPS

The design of rotary gear pumps consists of two or more gears meshing, which are engaged in a closely fitted housing. Gear pumps normally have a flow rate of around 70 L/min (18.5 gpm) and delivery pressure up to 220 bar (3,190 psi). The gear pumps can be classified into the following types:

1. External gear pump

2. Planetary or internal gear pump

Gear pumps are classified into external gear type and internal gear type. Although external gear pumps have been used most widely in the past, internal gear pumps are increasingly gained popular in recent years as more users demand for low noise and pulsation.

External Gear Pumps

External gear pumps are designed with two-gear combinations, one of the gears is mounted on the drive shaft while the second gear is attached with the driven shaft. The gears are designed to rotate in opposite directions and mesh at a point in the housing between the inlet and outlet ports. The pumping action of the external gear pump is caused by the rotation of the two gears.

As external gear pumps are of relatively simple in construction, it can be easily worked, along with high durability and acceptable tolerance against dust than other types. They are widely used in construction equipment and industrial machineries.

Internal Gear Pumps

The internal gear is a modification of the external gear pump and also uses two set of gears. The spur gear is mounted inside a larger ring where the smaller spur gear is in mesh with one side of the larger internal gear. It is kept apart by a separator on the other side. As in the external gear pumps, the fluid moves from the suction port to the discharge port by entrapment action between the meshed teeth of the rotating gears. Input energy can be applied either to the inner ring gear or to the outer ring gear. It is also to be noted that the direction of rotation of both gears is the same.

In the past, internal gear pumps were chiefly used as low pressure pumps. However, with improvement in design construction and technology advancement, high pressure internal gear pumps have been developed. The advantage of small pulsation in discharge and low noise generation, they became very popular as it offers a wide range of pressure. They are used in high pressure machinery such as injection moulding machines, forging machines and medium pressure machineries including machining tools.

One other feature is it compact in size as compared to other pumps. They are commonly used as oil pumps or power steering systems for motor vehicles.

S/N	Descriptions	Physical size	Section view of pump
1.	External Gear		
2.	Internal Gear		
3.	Hydraulic Symbol		

Figure 2.4

Another form of the internal gear pump is the 'Gerotor' pump. Gerotor pumps are one of the most common types of internal gear pumps whose operation is quite similar to that of an internal gear pump. The inner gear rotor (gerotor element) is power driven and draws the outer gear rotor around as they mesh together. This forms the inlet and outlet discharge pumping chambers between the rotor lobes. The tips of the inner and the outer lobes make contact to seal the pumping chambers from each other. The inner gear has one tooth less than the outer gear, and the volumetric displacement is determined by the space formed by the extra tooth in the outer gear.

PISTON PUMPS

There are 2 main types of piston pumps, namely

1. Axial piston pump

2. Radial piston pump

S/N	Descriptions	Physical appearance	Sectional View
1.1	Axial piston pump		
1.2	Axial piston pump (bent)		
2.1	Radial piston pump – type 1		
2.2	Radial piston pump – type 2		
3.	Hydraulic Symbols		
	NB: An arrow across denotes variable pump types		

Figure 2.5

The basic working principle of piston pump is quite similar to a typical air cylinder bicycle pump except that in the hydraulic piston pump, there are about 6 to 8 small pistons (each size of about 10 mL syringe pump) reciprocating within the pump casing. The reciprocating action causes the suction of hydraulic oil and discharges it at very high pressure through the outlet port.

Like the gear pump, all piston pumps also have an inlet port and outlet port. The pressure of piston pump can reach very high pressure. Usually piston pumps are selected for high pressure applications and occasionally the combination of both pressure and flow rate. These pumps have vast applications in many areas especially in heavy duty industries, like in marine and shipbuilding and construction equipment.

HAND PUMPS

Hand pump is a mechanical device to pump hydraulic oil to deliver it with high pressure in one direction as in figure 2.6.

S/N	Descriptions	Physical Appearance	Sectional view
1.	Hand Pump		
2.	Hydraulic Symbol		

Figure 2.6

It consists of a suction unit where oil is allowed into a chamber in one direction only. This chamber with a movable piston set has an inlet port to receive oil and an outlet port to deliver the oil at high pressure. This is made possible with two non-return check valves in it construction.

Pump Performances

The performance of a pump is primarily a function of the precision of its manufacturing. Components made to close tolerances should be maintained while the pump is operating under specific design conditions. Maintenance of close tolerances is achieved by designing system which incorporate both mechanical integrity and balanced pressures.

Theoretically, an ideal pump is one with zero clearances between all mating parts. Although this is not feasible from the design and manufacturing point of view, the working tolerances should be normally be made as small as possible.

Pump manufacturers run tests to determine the performance of their various types of pumps. Overall efficiency of the pumps can be computed by comparing the output of the pump to the power supplied at the input.

Pump Efficiencies

The performance data for pumps is determined by a series of tests usually carried out by manufacturers. By comparing the actual hydraulic power output of a pump with the mechanical input comparing the actual hydraulic power output of a pump with the mechanical input power supplied by the prime mover, its overall efficiency can be computed. The overall efficiency can in turn be broken into two distinct components namely, volumetric efficiency and mechanical efficiency.

Volumetric efficiency

This indicates the amount of leakage. It takes place within the pump and involves considerations such as manufacturing tolerances and flexing of the pump casing.

Pump capacity

Pump manufacturers, usually specify the pump performance characteristics and performance curves with graphs, for better visual interpretation. The variation in actual pump capacity depends mainly on the following three factors:

1. Discharge pressure: The higher the discharge pressure, greater the internal leakage hence lower is the actual capacity.
2. Running clearances: Large clearances mean greater internal leakage.
3. Oil viscosity: The use of low-viscosity oil leads to greater internal leakages.

To keep the design consideration common globally, all gear pumps are designed for their rated capacity at a certain constant pressure. When the pressure at the discharge of a pump increases, the flow rate of the pump reduces. Hence, while designing a pumping system, care should be taken to ensure that the discharge line offers the least resistance to pump discharge.

Effect of running clearance on capacity

Clearances exceeding specified tolerance levels tend to adversely affect the pump performance and efficiency through increased leakage. Let us analyse this further, specifically in relation to gear pumps, in order to examine how any dilution in running clearances end up adversely affecting pump performance.

All gear pumps are manufactured precisely with specific minimum clearances as per design. However, it is not possible to actually have a zero clearance. Phenomena such as wear, scuffing, abrasive friction and rusting are said to be the major causes for increase in this clearance. Investigation has found that a 0.04 mm side clearance produces eight times as much internal leakage compared to a 0.02mm side clearance.

Tests were conducted on the running clearance with 1,500 rpm pump speed. It was then intentionally varied, to determine its effect on the pump capacity. During the course of these tests the following findings were established.

A change in the side clearance from 0.025 to 0.045 mm resulted in a 20% reduction in the pump capacity. The pump capacity was found to reduce further with a decrease in pump speed from 1,500 to 400 rpm with the other conditions being held constant.

Effect of oil viscosity on capacity

If the viscosity of the oil is high, the pump capacity is found to increase and vice versa. Since the viscosity is a function of temperature, cold oil results in a higher capacity.

For a similar pump as used earlier and for the operating conditions listed below, the following observations were recorded.

1. A 20 °C (68 °F) rise in oil temperature led to a decrease in pump capacity by 20%.

2. Upon further lowering of the pump speed, the pump capacity was found to decrease further.

It is therefore clear that the pump capacity is closely related to the temperature of the oil being used. Hence, it is quite imperative for any testing procedure involving the determination of the pump internal condition or its capacity, to specify the oil temperature, otherwise the whole exercise is meaningless.

Noise

Noise is another important parameters used to determine pump performance. It is measured in unit of decibels (dB). Any increase in the noise level normally indicates wear and tear, which may result in pump failure. However, the noise we hear from the pump in operation is not just the sound coming from the pump, but it includes vibration and the fluid pulsations as well. In general, fixed displacement pumps are less noisier than variable displacement pumps as they have a rigid construction.

Pump speed has a strong influence on its noise, while the size and pressure have about an equal or lesser effect. Another common cause for noise in hydraulic system is the presence of entrapped air bubbles in the fluid. These air bubbles, even if they represent less than 1% by volume, it will change the compressibility of the fluid so much that they can sometimes cause even a fairly silent pump to operate with excessive noise. Often, this leads to cavitations in pumps and sometimes can be very serious.

Sometimes, you may be required to check noise level of pump in certain installations. The procedure can be easily administered quickly with simple noise level meter. If you did not have one with you, a smart phone could also render a quick check. Firstly of course, you have to **download an app (it is free)** to your smart phone to measure noise level. It is quite easy and fairly accurate. (NB: Try it. You will find it is astonishingly accurate)

Measuring conditions

Viscosity of oil used: 20 cSt (Oil equivalent to LSO VG32, oil temperature at 50° C, or 122 °F)

Table 2.2 Noise characteristics (An example)

Measuring Conditions	Observation & Findings
Viscosity of oil used	20 cSt
Measuring location	1m behind pump
Background noise	40 dBA

Cavitation in pumps

Pump cavitation is another phenomenon associated with noise and caused by the presence of entrained air bubbles in the hydraulic fluid or vaporization of the hydraulic fluid. It normally occurs when the pump suction lift is excessive and the pump inlet pressure falls below the vapour pressure of the fluid. As a result of this, the air bubbles that originate in the low-pressure inlet side of the pump collapse upon reaching the high-pressure discharge side, resulting in increased fluid velocity and giving rise to impact forces which can erode the pump components and shorten its life.

Following are a list of measures that can be undertaken to address and minimize the effect of cavitation:

1. Maintaining short pump inlet lines.

2. Maintaining suction line velocities below 1.5 m/s (4.9 ft/s).

3. Positioning of the pump as close to the reservoir as possible.

4. Minimizing the number of inlet line fittings.

5. Using low-pressure drop filters and strainers in the inlet side.

6. Using the proper oil according to the recommendation of the manufacturer.

Type of Pumps comparison

The table below shows the typical pump specifications using both pressure rating and displacement to perform sizing for practical usage in any piece of hydraulic power equipment. For example, there is no difference in physical sizes between a 200 bar and a 350 bar gear pump with similar discharge rate.

Table 2.2 and 2.3 – Commercial available pump types and their performance parameters.

These are typical pump designs in the market. Please note that some manufacturers may have difference designs and physical appearances, but basically they falls into these three categories as tabulated in Table 2.2 and 2.3 below

Table 2.2 - Pumps and capacity

Pump Type	Pressure (kgf/cm^2)	Discharge (l/min)	Maximum Speed (rpm)	Overall efficiency (%)
Gear	20 ~ 175	7 ~ 750	1,800 ~ 7,000	75 ~ 90
Vane	20 ~ 175	2 ~ 950	2,000 ~ 4,000	75 ~ 92
Axial piston	70 ~ 350	2 ~ 1,700	600 ~ 6,000	85 ~ 98
Radial piston	50 ~ 250	20 ~ 700	700 ~ 1,800	80 ~ 92
Hand Pump	Up to 350	Discharge by stroke		
	NB: Discharge capacity usually determine the physical size of pumps.			

Table 2.3 - Pumps and their major characteristics

Major Features	Vane Pump	Gear Pump	Piston Pump
Working Principle	Varying volume between vane & cam ring	Shifting volume surrounded by gear grooves & casing	Varying volume by reciprocating pistons
Ave Efficiency	High	Low except the pressure loading type	Highest among the three types
Viscosity & efficiency	Significantly affected	Most significantly affected	Not affected
Wear & efficiency	Does not decrease since wearing of cam ring or vane can be compensated	Decreases with wearing	Decreases with wearing
Sensitivity to dust	Sensitive to relatively small dust particles, but less affected than the piston type	Low pressure type is less affected, while the pressure loading type is sensitive	Most sensitive since components are assembled with small clearance
Variable Displacement	Possible	Not possible	Possible

Pump Selection Criteria

Pumps are selected for a particular application in a hydraulic system based on a number of factors some of which are as follows:

1. Flow rate requirement
2. Operating speed
3. Pressure rating
4. Performance
5. Reliability
6. Maintenance
7. Cost and
8. Noise

The selection of a pump typically entails the following sequence of operations:

1. Selection of the appropriate actuators (cylinder or motor) based on the load.
2. Determining the flow-rate requirements, this involves a calculation to determine the flow rate required to drive the actuator through a specified distance, within a given time limit.
3. Determination of the pump speed and selection of the prime mover, this together with the flow rate computed as in '2' helps to determine the pump size (volumetric displacement).
4. Selection of the pump-type based on the application.
5. Selection of system pressure requirements, this also involves determination of the total power to be delivered by the pump.

NB: Refer to Chapter 7 for details of above procedures, step-by-step calculations, sizing and selection of hydraulic components.

Practical Situation on Site

Site Identification and What to check if you are a hydraulic Engineer?

Followings are some hand on procedures to consider, if you happen to attend to a hydraulic pump breakdown on site. There are few steps you can do.

1. Examine the name plate of the pump. It would normally bear the manufacturer information, type of pumps, pressure and flow rate information usually in cc/rev.

2. Request to look at the operation manual with the hydraulic circuit and pump specifications would be sufficient to start your checklist for further actions. Consult the operator (the one who know about the system).

3. If the pump is a small one, chances that it is either a gear or vane pump. For piston pump, most often the physical shape itself is sufficiently good indication.

4. If all of the above are not available to you. There are two other sensible steps you could do.

a. Check as much information of the load ends application, the tank size, sizes of hoses and pipes, working pressure (pressure gauge), note the environmental factors and make a rough sketch of the hydraulic circuit of the system. A lot of times you can be right about the pump type and it system. Of course, the easier solution is one to one replacement of the same pump type and size. That can be easily done with.

b. If you have full rights to the whole system. Then you could isolate the pump set and the loads or actuators. Be very careful about dismounting the pump for service or replacement because there are 2 typical HPU designs, namely the horizontal mount and vertical mount. For horizontal mount pump set, it is seated on tank top cover or by the tank side on structural skid as in Figure 2.7 (1). It is visible and easy to isolate. For the vertical mount pump set as in Figure 2.7 (3), it is below the tank top cover. It is not visible and isolating it from the load is not as easy without lifting the whole top cover. Most hydraulic power units (HPUs) are equipped with shut off valves and isolating valves separating the HPU and the various types of actuators.

It is always a good practice to inspect the pressure gauge for present of pressure in the system before performing anything else, even though electrical power is switched off.

S/N	Descriptions	Physical Appearance
1.	Horizontal HPU	
2.	Horizontal HPU	
3.	Vertical HPU	

Figure. 2.7

Some HPUs may have accumulators as shown in Figure 2.7 (1) that store excessive pressure and these can be easily isolated with needle valves. These valves are usually fitted together with the accumulators. Anyway, whatever you are doing, be very careful with hydraulic machines.

NB: Respect the power of hydraulic machines. Though it may be an idle piece of equipment with electrical power 'off' but do not be deceived by it.

Chapter 3 – Flow Control Valves

Control Valves
Check Valves (Non return & pilot operated)
Pressure Control Valves
Pressure Relief Valves
Pressure Reducing Valves
Unloading Valves
Sequencing Valves
Counterbalance Valves
Brake Valves
Directional Control Valves
Classification of Valves by Spool Designs
Three-way & Four-way Control Valves
Solenoid Directional Control Valves
Manual Directional Control Valves
Servo Control Valves
Flow Control Valves (Meter-in & Meter-out)
Shuttle Valves
Flow Divider valves
Stackable Valves System (CETOP/NFPA)
On/Off Valves
Manifolds

Upon completion of this chapter, you should be able to:

1. Recognise the most common flow control valves
2. Recognise the hydraulic symbols of each control valves
3. The various types flow control and its applications
4. Know about why the particular control valves are being used
5. Recognise the type of flow control valves used on site
6. Present and sketch the basic hydraulic circuit diagram with the various control valves

You cannot change what has already happened. You can always change the way you respond
Anthony Robbins

Control Valves

A control valve is a device used for adjusting or manipulating the flow rate of a liquid or gas in a pipeline. The valve essentially consists of a flow passage whose flow area can be varied. The external motion can manipulate either manually or from an actuator positioned pneumatically, electrically or hydraulically. In response to some external positioning signal. This combination of the valve and actuator is known as a control valve or an automatic control valve.

Basically, there are three types of control valves,

1. Direction control valves: Direction control valves determine the path through which a fluid traverses within a given circuit. In other words, these valves are used to control the direction of flow in a hydraulic circuit. It is that component of a hydraulic system that starts, stops and changes the direction of the fluid flow. Generally, the direction control valve actually designates the type of hydraulic system designs, either open or closed. An example of their application in a hydraulic system is the actuator circuit, where they establish the direction of motion of a hydraulic cylinder or a motor.

2. Pressure control valves: Pressure control valves protect the system against overpressure conditions that may occur either on account of a gradual build up due to decrease in fluid demand or a sudden surge due to opening or closing of the valves. Pressure relief, pressure reducing, sequencing, unloading, brake and counterbalance valves control the gradual build-up of pressure in a hydraulic system. Pressure surges can produce instantaneous increases in pressure as much as four times the normal system pressure and that is the reason why pressure control devices are a must in any hydraulic circuit. Hydraulic devices such as shock absorbers are designed to smoothen the pressure surges and also to dampen hydraulic shock.

3. Flow control valves: The fluid flow rate in a hydraulic system is controlled by flow control valves. Flow control valves regulate the volume of oil supplied to different parts of a hydraulic system. Non-compressed flow control valves are used where precise speed control is not required, since the flow rate varies with the pressure drop across a flow control valve. Pressure-compensated flow control valves are used in order to produce a constant flow rate. These valves have the tendency to automatically adjust to changes in pressure..

Since it is important to know the primary function and operation of the various types of control components, it is required to examine and study the functioning of each of these valves in detail.

In order to help understand the various functions of each valve, a simple hydraulic circuit diagram as depicted in Figure 3.1, is always helpful. It is advisable to refer to the diagram as and when require. At some point as we move along or whenever appropriate, certain visualization skill would be sufficient to engage this topic.

Check Valves

Figure 3.1 shows the symbolic representation of a check valve along with a simple check valve application in an accumulator circuit.

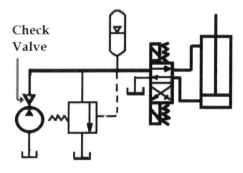

Figure. 3.1

Check valve

Figure 3.1 shows the symbolic representation of a check valve along with a simple check valve application in an accumulator circuit. As the name implies, direction control valves are used to control the direction of flow in a hydraulic circuit. The simplest type is a check valve, which is a one-way direction control valve. It is a one-way valve because it permits free flow in one direction and prevents any flow in the opposite direction.

Figure 3.2 shows the internal operation of a check valve. As you can see, a light spring holds the poppet in the closed position. In the free flow direction, the fluid pressure overcomes the spring force. If the flow is attempted in the opposite direction, the fluid pressure pushes the poppet (along with spring force) in the closed position. Therefore, no flow is permitted. The higher the pressure, the greater will be the force pushing the poppet against its seat.

Check Valves (Non-Return Check Valves)

S/N	Descriptions	Physical Appearance	Sectional view
1.	Check Valve (Non-return Valve)		
2.	Hydraulic Symbol		

Figure 3.2

Check valves or sometimes known as non-return check valves are used to regulate flow in hydraulic circuits to a certain direction. It is designed to open by a specified cracking pressure in one direction while cutting off the flow completely in the opposite direction. Check valves are classified into inline type and angle type.

It is common term to describe hydraulic check valve with it cracking pressure. Cracking pressure means that a minimum pressure is required for oil to lift the ball seat (pressed on the v-via by a compression spring force) to open the valve and oil could then flow through. The flow resistance varies with respect to the increase in flow rate; this is because the flow has to overcome the spring force which constantly acts to shut it.

Pilot-operated check valve

The second type of check valve is the pilot operated check valve. The cross-section of a typical pilot operated check valve has been illustrated in Figure 3.3

S/N	Descriptions	Physical appearance	Sectional view
1.	Pilot Operated Check Valve		
2.	Hydraulic Symbol		

Figure 3.3

This type of check valve always enables free flow in one direction but permits flow in the normally blocked opposite direction also if the pilot pressure is applied at the pilot pressure port of the valve. The check valve poppet has a pilot piston attached to the threaded poppet stem by a nut. The light spring holds the poppet seated in a no-flow condition by pushing against the pilot piston. The purpose of the separate drain port is to prevent oil from creating a pressure build up at the bottom of the piston.

Pilot check valves are often used in hydraulic systems where it is desirable to stop the check action of the valve for a portion of the equipment cycle. An example of its application with its use is found in the control situation where you want to lock the hydraulic cylinders in position.

Pilot controlled check valves

Pilot controlled check valves can move pilot pistons by remote pressure to push up closed check valves for reverse flow. They are divided into general type and de-compression type. The de-compression type incorporates the de-compression poppet to the main poppet. Pilot pressure pushes up the pilot piston to open the de-compression poppet and then, after the pressure decreases, open the main poppet.

Pressure Control Valves

We have already briefly discussed about pressure control valves and their basic functions in a hydraulic system. This section is to give a deeper understanding of the concept on manipulating the force through a hydraulic system using pressure control valves. Simple examples are also illustrated along the way to showcase the operating and design principles of the various types of pressure control valves and their applications. The two basic pressure control valve design types are:

1. Direct-acting pressure control valves and
2. Pilot-operated pressure control valve

The operating principles of all the pressure control valves revolve around with these two basic design types.

The primary objective in any hydraulic circuit is to either control the flow rate or pressure. There are six different types of pressure control valves that have been developed for accurate control of force in a hydraulic circuit. These are illustrated in the figures below with their hydraulic symbols.

One may find the symbols quite confusing since these valves resemble one another so closely, almost identical physically. Often times, only their location of use in a hydraulic circuit may actually help to determine the type of pressure control valves.

Pressure control valves are designed to control pressure in oil hydraulic circuits and are classified according to their purposes as follows.

Pressure Relief valves
(Also known as Pressure Setting Valves)

The most widely used type of pressure control valve is the pressure relief valve since it is found in practically every hydraulic system. It is normally closed valve whose function is to limit the pressure to a specified maximum value by diverting the pump flow back to the tank. The primary port of a relief valve is connected to system pressure and the secondary port connected to the tank.

When the relief valve's poppet is actuated at a predetermined pressure, a connection is established between the primary and secondary ports resulting in the flow being diverted to the tank.

A poppet is held seated inside the valve by the direct force of a mechanical spring which is usually adjustable. The poppet is kept closed by the spring tension set on the knob until the system pressure working against the poppet reaches the cracking pressure.

The poppet is forced off its seat when the system pressure reaches full relief value. This permits fluid flow across the poppet to the tank. Thus the required pressure in the system is maintained as per the pre-set value on the pressure relief valve.

When the hydraulic system does not accept any flow due to safety reason in the system, the pressure relief valve releases the fluid back to the tank. As such, it maintains the desired system pressure in the hydraulic circuit. At the same time, it also provides protection against any overloads experienced by the actuators in the hydraulic system. One other important function of a pressure relief valve is to limit the force or torque produced by the hydraulic cylinders and motors.

Another important consideration to be taken note of is, the practical difficulty in designing a relief valve spring strong enough to keep the poppet closed at high-flow and high pressure conditions. This is normally the reason why direct acting relief valves are available only in relatively smaller capacities.

S/N	Descriptions	Physical Appearance	Sectional View
1.	Direct Operated (PRV)		
2.	Balance Piston type (PRV)		
3.	Hydraulic Symbol		

Figure 3.4

Pressure relief valves are a compulsory valve for every hydraulic circuit systems, because this valve also set the system pressure that the pump will drive. Relief valves, as it term implies, are used to protect the pumps and all the other control valves from excessive pressure. The result of over excessive pressure in the hydraulic system will cause premature damage to pressure sensitive parts located downstream like, seals, o-rings, fittings and pipe joints.

There are two main types of pressure relief valves. They are direct operated types which only handle small flow system and the balance piston types are used to handle higher flow capacity.

Direct operated types of valves are of simple construction as shown in picture above. It is compact in size. A spring loaded spindle-like configuration with set of o-rings mounted on a solid cartridge with an inlet and outlet ports. Ports are threaded with either NPT or BSP. Occasionally, you would find some PRV with SAE ports. It can handle relatively large pressure override (meaning pressure setting range) and often cause chattering phenomenon. This type of valve is very common in many hydraulic applications and usually handles relatively small flow.

Pressure-reducing valves

Pressure-reducing valves are normally open pressure control valves that are used to limit pressure in some circumstances of a hydraulic circuit. Reduced pressure results in a reduced force being generated. This is the only pressure control valve which is of the normally open type. A typical pressure reducing valve and its function is described below.

S/N	Descriptions	Physical Appearance	Sectional View
1.	Pressure-reducing valve		
2.	Hydraulic Symbol	REDUCING VALVE	

Figure 3.5

This valve is actuated by the downstream pressure and tends to close as the pressure reaches the valve setting. When the downstream pressure is below the valve setting, fluid will flow freely from the inlet to the outlet. Observe that there is an internal passage from the outlet, which transmits the outlet pressure to the spool end opposite the spring.

When the downstream pressure increases beyond the value of the spring setting, the spool moves to the right to partially block the outlet port as shown in sectional view of Figure 3.5. Just enough flow is thus passed through the outlet to maintain its preset pressure. If the valve closes completely, leakage past the spool could cause the downstream pressure to build up above the set pressure of the spring. This is prevented from occurring by allowing a continuous bleeding to the tank through a separate drain line.

Unloading Valves

Unloading valves are remotely piloted, normally closed pressure control valves, used to direct flow to the tank when pressure at a particular location in a hydraulic circuit reaches a predetermined value. **Figure 3.6** depicts the sectional view of a typical unloading valve used in hydraulic systems.

S/N	Descriptions	Physical Appearance	Sectional View
1.	Unloading Valve		UNLOADING SPOOL / B-(T) / X / A / CONTROL ORIFICE
2.	Hydraulic Symbol		UNLOADING VALVE

Figure 3.6

The unloading valve is used to unload pressure from the pump connected to port A, when the pressure at port X is maintained at a value satisfying the valve setting. The spring-loaded ball exercises control over the high-flow poppet along with the pressure applied at port X. Flow entering at port A is blocked by the poppet at low pressures. The pressure signal from port A passes through the orifice in the main poppet to the top side area and then to the ball. There is no flow through these sections of the valve until the pressure rise equals the maximum value permitted by the spring-loaded ball. When that occurs, the poppet lifts causing fluid flow from port A to port B which in turn is connected to the tank. The pressure signal at port X acts against the solid control piston and forces the ball further off the seat. Due to this, the topside pressure on the main poppet reduces and allows flow from port A to B with a very low-pressure drop, as long as the signal pressure at port X is maintained.

Sequencing Valves

A sequencing valve again is a normally closed pressure control valve used for ensuring a sequential operation in a hydraulic circuit, based on pressure. In other words, sequencing valves ensure the occurrence of one operation before the other. A sectional view of a sequencing valve is shown in Figure 3.7.

When the components connected to port A of the valve reach the pressure set on the valve, the fluid is passed by the valve through port B to do additional work in a different portion of the system. The high-flow poppet of the sequence valve is controlled by the spring-loaded cone. At low pressure, the poppet blocks the flow of fluid from entering port A.

The pressure signal at port A passes through the orifices to the top side of the poppet and to the cone. There is no flow through the valve unless the pressure at port A exceeds the maximum set pressure on the spring-loaded cone. When the pressure reaches the set valve, the main poppet lifts, allowing the flow to pass through port B.

It maintains the adjusted pressure at port A until the pressure at port B rises to the same value. A small pilot flow (about ¼ gpm or

0.9 L/min) goes through the control piston and past the pilot cone to the external drain.

S/N	Descriptions	Physical Appearance	Sectional View
1.	Sequencing Valve		A B
2.	Hydraulic Symbol	SEQUENCE VALVE	

Figure 3.7

When there is subsequent pressure increase in port B, the control piston acts to prevent further pilot flow loss. The main poppet opens fully and allows the pressures at port A and B to rise together. Flow may go either way during this condition.

Counterbalance Valves

A counterbalance valves are also a type of normally closed pressure control valve and is particularly used in cylinder applications for countering a weight or overrunning load. Figure below shows the operation of a typical example of counterbalance valve mounted to the side of a hydraulic cylinder.

S/N	Descriptions	Physical Appearance	Sectional View
1.	Counterbalance Valve		
2.	Hydraulic Symbol	COUNTERBALANCE VALVE	

Figure 3.8

The primary port of this valve is connected to the lower port of the cylinder (Figure 3.8a) and the secondary port is connected to the direction control valve. The pressure setting of the counterbalance valve is kept higher than required to prevent the cylinder load from falling.

When the pump flow is directed to the top of the cylinder through the DCV, the cylinder piston is pushed downward. This causes the pressure at the primary port to increase and raise the spool. This results in the opening of a flow path for discharge through the secondary port to the DCV and back to the tank.

When raising the cylinder, an integral check valve opens to allow free flow for retraction of the cylinder, Figure 3.8a is an illustration of how exactly the counterbalance valve operates in a hydraulic circuit. As shown in the figure, the counterbalance valve is placed just after the cylinder in order to avoid any uncontrolled operation.

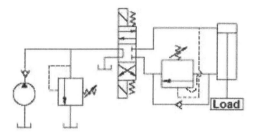

Fig.3.8a Example using counterbalance valve

However, if the counterbalance valve is not provided in the above example, there would be an uncontrolled fall of the load. This would happen because the pump could not keep pace with the flow. The counterbalance valve is set to a pressure slightly higher than the load-induced pressure. As the cylinder is extended, there must be a slight increase in pressure in order to drive the load down.

Brake Valves

Brake valves are normally closed pressure control valves that are frequently used with hydraulic motor for dynamic braking. The operation of these valves involves both direct and remote pilots connected simultaneously. During operation, the valve is kept open through remote piloting, using system pressure. This results in eliminating any back pressure on the motor that might cause downstream resistance and subsequent load on the motor.

When the direction control valve is de-energized, remote pilot pressure is lost allowing the valve to close. The valve is then driven open through the internal pilot, by the inertia of the load, resulting in dynamic braking.

Directional Control Valves (DCV)

As briefly discussed above, direction control valves are used to control the direction of flow in a hydraulic circuit. They are primarily classified by their number of possible positions, port connections or ways and the manner in which they are actuated. For example, the number of porting connections is classified as ways or possible flow paths. A four-way valve would comprise of four ports **P, T, A and B**. Using diagrams and symbols for hydraulic configuration make learning and understanding hydraulic system easier. For example, a three-position valve is indicated by rectangle boxes with 'Arrows' and 'Ts' as shown in tables below. There are several mechanisms employed for actuation or shifting of the valve. They include hand lever, foot pedal, push button, mechanical, hydraulic pilot, air pilot, solenoid and spring.

NB: Note the flow directions of arrows and 'T' block (**NB: Do not confuse this 'T' block to that of 'T' for tank or 'T' ti-ed joint in this and later chapters**).

Directional Control Valves

As mentioned earlier, directional control valves are designed to start, stop, control direction, accelerate, and decelerate cylinders and hydraulic motors. They are used in a variety of applications and available in many types.

1. Classification of directional control valves

2. Classification according to number of ports and control positions

In directional control valves, the number of ports means the number of pipes connected, and the number of control positions means the number of valve positions.

In addition to forward or reverse operations of hydraulic motors, directions control valves must have a function to feed the oil hydraulic system to specifications and purposes of actuators when stopping them (when the directional control valve is at neutral position).

For instance, if all ports of 3 position valve are closed when the valve is at neutral position, cylinders will be locked at the position and pressure on the pump side is maintained at a pre-set level for relief valve, allowing other systems or actuators to operate. In this case, if the center bypass valve, allowing no load operation is possible on the pump circuit side while the cylinders are locked in position to reduce heat generation. This arrangement facilitates power saving and increase pump's service life. At the same time, however, care should be taken for connection method if other systems are operated, since no pressure is generated in these systems.

The above description, only serve to illustrate a particular control valve function. In hydraulic controls and systems, it is not just one type of valve fit all control purposes. For beginners, those symbols can be very confusing and discouraging to learn and understand them. So, in order to understand the various types of DCV and its function. It is easier to visualize the notation of the valves' symbol by firstly, examining the ports and positions configurations as depicted in Table 3.1 then reading the spools' type and spring function in Table 3.2 and Table 3.3 will make more sense to you.

These configurations are some common symbols which you will find them labelled on type plate of each valve to describe the valve's functions.

Table 3.1 - Ports and Positions Representation

Classification		Symbol	Descriptions
Nr. of Port (connection)	2 - Port		With 2 connection ports. Use to open or close a line
	3 - Port		With 3 connection ports. Use for controlling flow from a pump to 2 directions
	4 - Port		With 4 connection ports. For forward, reverse and stopping operations.
	Multiple - port		With 5 or more connection ports. Use for special purposes.
Nr. of Control Positions	2-Position		With 2 control positions.
	3-position		With 3 control positions.
	Multiple-position		With 4 or more control positions. For special purposes.

Classification of Valves by Spool Designs

In addition to functional classification, directional control valves are typically group under 4 major spool types according to type of operations. There are other more complex spool designs but for now knowing and understanding these four types are purpose of this book.

Table 3.2 - Spools Feature

S/N	Spool	Symbol	Descriptions
1.	Close Center	(A B / P T)	All ports are closed at neutral position. When operation, surge pressure is generated during the cross-over.
2.	Open Center	(A B / P T)	All ports are connected at neutral position. Commonly use to unload the pump and set cylinders floating at neutral position
3.	Center Bypass	(A B / P T)	Only A and B ports are closed and the pump is unloaded at neutral. Commonly use with series connecting 2 or more units in the system
4.	ABT connection	(A B / P T)	Only pump port is closed at neutral position.

Mode of operations on the various DCVs

In term of operation mode of valves types, they are classified under 3 main groups namely,

1. Manually Operated type,
2. Mechanically Actuated or Operated by hydraulic or pneumatic piloted type and
3. Electromagnetic actuated commonly known as 'Solenoid'.

NB: Note also there are valves designed with or without spring functions. Non-spring version includes the detent model which is capable of holding valve position which will be explained in details later.

Table 3.3 - Common Valves' Configurations

S/N	Classification	Symbol	Descriptions
1.	Manual		Operate by lever.
2.	Mechanical		Operated by cam, roller and other mechanical means.
3.	Pilot pressure (air or oil)		Operated by pilot oil hydraulic or pneumatic.
4.	Solenoid		Control by electromagnetic force (solenoid).
5.	Solenoid - hydraulic pilot		Main spool valve operated by hydraulic pilot activated by solenoid.
6.	Spring offset		Directional control by solenoid 'ON' and return to original position by spring when solenoid 'OFF'.
7.	Detent Type (No spring)		In detent type of valve, position is held in place even with the present of vibration.
8.	Spring Center		Spool returns to original position by spring action when solenoid 'OFF'.

Constructions and features of directional control valves

Normally open and normally closed

Direction control valves may also be categorized as normally open and normally closed valves. This terminology would normally accompany the direction control valves, as reflected in the above examples of two-position valves. We shall briefly discussed them below.

Open and closed center hydraulic circuits

For most control valves system, the majority of the hydraulic circuits are basically categorized into two main types of design namely, open center and closed center. The type of circuit is usually designated by direction control valves. In open center circuits, the pump flow is routed back to the tank through the direction control valve during neutral or dwell time.

Normally, in this type of circuit, a fixed displacement pump such as gear pump is used. In the event of flow blocked during neutral or the direction control valve being centered, flow tends to get forced over the relief valve, possibly resulted in excessive amount of heat.

In the closed center circuit, the pump flow is blocked at the direction control valve both in neutral or when the valve is centered. In this case, either a pressure-compensated pump such as a piston pump that de-strokes or an unloading circuit with a fixed displacement pump is used.

A neutral or central position is provided in a three-position direction control valve. This determines whether the circuit is open or closed and also the type of work application depending on the inter-connection between the P and T ports and the configuration of the A and B ports respectively. The four commonly used three-position direction control valves: the open type, the closed type, the tandem type and the flow type are illustrated in table 3.2 above.

In the tandem type, port P is connected to port T and both the port A and B are blocked. This results in an open circuit. This type of valve finds application in circuits involving fixed volume pumps where ports A and B being blocked, the load can still be held when the valve is at neutral position.

Discussion of Open Center and closed center operation

Open Center – principal of operation

An open center type of design means, when a hydraulic spool is at idle position, the valve is at 'off' position. If any oil delivery flows into it via the pump, the oil is allowed to pass directly through the spool back to the hydraulic reservoir (In short it means P to T, the directional control valve in Figure 3.8a is an example of an open center).

Advantages of Open center system

More cost effective to build
More cost effective to maintain
Easier to maintain
More tolerant to contaminants
Hydraulic pressure resists water infiltration

Closed Center principal of operation

When a hydraulic spool is idle, a closed center hydraulic circuit allows the oil flow to stop rather than return the oil to the hydraulic reservoir. Closed center circuits provide a spool with full power when the spool is engaged again. (In short, it means P & T block, the directional control valve in Figure 3.9 is an example of a closed center)

Typically closed center applications are topside applications. Topside means that the valve is seated on sub-plate or stack right on the top cover of power unit reservoir. It is very obvious and prominently located on one side of the hydraulic power unit. The pressure line is usually hidden below the cover running from the pump pressure side port to the underside of the sub-plate.

The sub-plate has 2 threaded port holes marked clearly with P (Pressure) and T (Tank) on the underside. On the opposite side (the top surface) is machine finished with 4 holes known as P, T and A, B where their respective locations are orientated by 4 threaded holes to receive the particular valve or valves mounting.

(Refer also to Figure 3.20 to examine the manifold blocks which has the similar holes orientation. It helps to visualize how a sub-plate would be like.)

Advantages of closed center system

More cost effective to operate
This system allows spool to be changed without having to shut down the hydraulic system.

Spool-type direction control valves

As discussed earlier, in spool-type direction control valves, spools incorporated in the control valve body are used to provide different flow paths. This is accomplished by the opening and closing of discrete ports by the spool lands. The spool is a cylindrical member which has large-diameter lands, machined to slide in a very close-fitting bore of the valve body.

The radial clearance is usually less than 0.02mm. The spools may be operated through different means like mechanical actuation, manual operation, pneumatic operation, hydraulic or pilot control and electrical operation.

Two-way directional valves

This type of directional valve is designed to allow flow in either direction between two ports. Figure 3.1 above showing a check valve application in an accumulator circuit is a typical example of a two-way, two-position on-off valve. Its function is to connect the accumulator to the load whenever desired. To put it rather simply, this valve is the hydraulic equivalent of a regular single-pole, single-throw (SPST) on-off electrical switch.

Three-way and Four-way Directional Control Valves

Another type of direction control valve is the three-way and four-way valve, containing three and four ports (as explained in the above tables) depicts the flow paths through two four-way valves. As shown in the figure, one of these valves is used as a three-way valve since its port T leading to the oil tank is blocked. One of the simplest ways by which a valve port could be blocked is by screwing a threaded plug into the port opening (NB: Ensure they are of the same thread types).

Referring to the hydraulic symbols, the flow entering at pump port 'P' (port connected to the pump discharge line) can be directed to either of the outlet ports 'A' an 'B'. Most direction control valves use a sliding spool to change the path of flow through the valve. For a given position of the spool, a unique flow path configuration exists within the valve. Directional valves are designed to operate with either two positions of the spool or three positions of the spool. Let us now analyse the flow paths in details.

Three-way valve (four-way valve used as a three-way valve)

Here the port **T** is blocked and only the other three ports **A, B** and **P** are used.

The flow can go through the valve in two unique ways depending on the spool position.

1. Spool position 1 : Flow can go from **P** to **B** as shown by the straight through line and arrow. Port **A** is blocked by the spool in this position.

2. Spool position 2 : Flow can go from **P** to **A**. Port **B** is blocked by the spool in this position.

Four-way valve

The flow can go through the valve in four unique ways depending on the spool position. Referring to the symbols,

1. Spool position 1 : Flow can go from **P** to **A** and **B** to **T**.
2. Spool position 2 : Flow can go from **A** to **T** and **P** to **B**.

The pump flow can be directed to either of two different parts of a circuit by a three-way directional valve. The typical application of a four-way directional valve in hydraulic circuits involves control of double-acting hydraulic cylinders. The positioning of the direction valve spool can be done manually, mechanically, by using electrical solenoids or by using pilot pressure.

Solenoid Directional Control Valves

The general principle of the solenoid action is very important in hydraulic system control. Like the relay and contactor, the solenoid is an electromechanical device. In this device, electrical energy is used to magnetically cause mechanical movement of a plunger.

(Refer to chapter 6 for further explanation about solenoid)

When the solenoid of this valve is energized, it resulted an electromagnetic force that pulls the armature of the coil into the magnetic field. As a result of this action, the connected push pin moves the spool in the same direction, while compressing the return spring. The shift in the spool valves results in port P opening to port A and port B opening to port T or tank, thereby allowing the cylinder to extend. When the coil is de-energized, the return springs move the spool back to its center position. Figure 3.9 shows a simple hydraulic circuit with a solenoid control valve (**P,T, A** and **B** ports blocked at neutral position) and Figure 3.10, the solenoid control valve (**P** to **T, A** and **B** ports blocked at neutral position).

A more detailed explanation of the valve's spool action is illustrated with the schematic steps in Table 3.4.

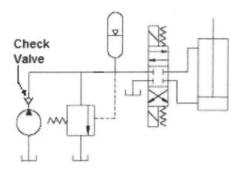

Figure 3.9

Solenoid-operated direction control valve

To further illustrate the working concept of solenoidal valves, a sectional view of double acting solenoid valve and its sequences are shown below in Table 3.4. In the de-energized condition as shown in Table 3.4 - (2) below, no pressure is available to either port **A** or port **B**, as it escapes around the land areas of the valve spool to the tank port. The centering springs keep the valve spool in this position, with both solenoids de-energized.

When solenoid **A** is energized, the force or pull exerted on the plunger in solenoid **A** moves the valve spool to the left. The spring in solenoid **A** is compressed (loaded). The land areas on the valve spool are now located so that pressure available at port P is free to flow into Port **A**. Port **B** is open to the tank port **T**. As there is generally little or no pressure connected at the tank port, it follows that there will be little or no pressure at port **B**.

When solenoid **B** is energized, the force or pull exerted on the plunger in solenoid **B** moves the valve spool to the right. The spring in solenoid **B** is compressed (loaded). The land areas on the valve spool are now located so that pressure available at port **P** is free to flow into port **B**. Port **A** is open to the tank port **T**. Oil will now discharge from port **A** through port **T**.

This valve has a flow capacity of 50 lpm and a maximum operating pressure of 250 kg/cm^2 (3,555 psi). It has a wet armature solenoid. The fluid around the armature serves to cool it and cushion its strokes without affecting the response time. There are no seals around the armature because of which its movement is not restricted. This allows all the power developed by the solenoid to be transmitted to the spool valve without the need to overcome seal friction.

S/N	Descriptions	Physical Appearance	Sectional view
1.	Solenoid Directional control valve		
2.	Symbol (**P-T, A & B** blocked)		

Figure 3.10

Table 3.4 - Spool Action

S/N	Sequence Of Spool Action and Oil Flow Directions (Spool type: P-T, A & B close)
1.	*[Symbol diagram: Solenoid A — A, B, P, T ports — Solenoid B]* This is the symbol of **P-T** with **A** and **B** ports close. The spring on both ends maintain the spool at neutral position. Solenoid of both ends are off.
2.	*[Diagram showing spool with A NO FLOW, B NO FLOW, P inlet, T outlet, X displacement]* Ports **A** and **B** are close. Spring force at both ends of spool are equal and lock at this position. 'X' is the spool displacement.

S/N	Table 3.4 - Spool Action (Continue)
3.	When a force **F** is energized on the left spool via solenoid, manual lever or air/oil pilot means. NB: oil/air pilot pressure is applied in opposite direction.
4.	The spool shifts to the left by 'X' distance opening port **A** and **B**. Now the oil flows from **P** to **A** and **B** to **T**. Spool will remain at this position as long as F_H, holding force is maintained.
5.	This figure is one of the common spool design for a flow type. Notice that the port recess is where oil will flow through and O-rings seats on **V**-grooves to form partitions & keep oil from flowing through the land.

Manual overrides are provided with most solenoid actuated valves, allowing for the spool to be operated by hand. This can be accomplished by depressing the pin in the push pin tube end located at each end of the valve.

Solenoid pilot directional control valves

Solenoid pilot directional control valves incorporate a solenoid directional control valve and hydraulic directional control valve. The solenoid directional control valve is operated as a pilot valve, and spool of main valve is operated by supplying pilot pressure.

Changeover time for these valves varies with pilot pressure.

Pilot controlled directional control valves

Pilot controlled directional control valves change over the spool by hydraulic pilot, and are used when separate installation of pilot directional control valve and main directional control valve is desired. These types of valves are chosen for design purposes, convenience purposes, remote situation and safety purposes (prohibiting electrical installation that). Usually the pilot pressure required for operating the spool is 5 to 7 kgf/cm^2 (71 to 99.5 psi). The pilot medium can be in the form of hydraulic oil or pneumatic (compressed air).

Direct acting and pilot-operated direction control valves

By direct acting, it is implied that some force is made to act directly on the spool, causing it to shift. A direct acting direction control valve can be actuated either manually or with the help of a solenoid as explained earlier.

For hydraulic systems that require higher flow rates above 132 lpm (35 gpm), a greater force is needed to shift the spool. The direct acting method cannot deliver the required force. It is not possible and therefore a pilot-operated arrangement is used. Figure 3.11 illustrates the working of a pilot-operated direction control valve. These valves are operated by applying oil (or air) pressure against a piston at either end of the valve spool. Referring to Figure 3.11 the top valve is known as the pilot valve, while the bottom valve is the main valve that is to be actuated. In most common pilot operated valve operation, the pilot valve is used to hydraulically actuate the main valve. The oil that use to power the piloting action on the pilot valve came either directly from the internal or an external source. Oil is directed to one side of the main spool, when the pilot valve is energized. The resultant shift in the spool leads to the opening of the pressure port to the work port thereby directing the return fluid back to the tank. External piloting or in other words, sending fluid to the pilot valve from an external source is often resorted to.

S/N	Descriptions	Physical Appearance	Sectional View
1.	Direct acting pilot operated DCV		
2.	Hydraulic Symbol		

Figure 3.11

The advantages associated with external piloting are that the effect of any other influence on the main system is not felt and the possibility of separate filtration ensures silt-free operation of the pilot valve. In addition, the valve may also be internally or externally drained. In the case of internal draining of the pilot valve, the oil flows directly into the tank chamber of the main valve. When operating the main control spool, pressure or flow surges occurring in the tank port may affect the unloaded side of the main valve as well as the pilot valve. This may be avoided by externally draining the pilot valve or in other words, feeding the pilot oil flow back to the tank.

To understand the concept of Pilot-operated direction control valves better, let us examine the sectional view of a pilot-operated four-way valve as shown in Figure 3.11.

The springs located at both ends of the spool push against the centering washers to center the spool when no oil is applied at the pilot ends. When oil (or air) pressure is supplied through the left passage, it pushes against the piston to shift the spool to the right. Similarly, when oil (or air) is introduced through the right passage, its pressure pushes against the piston to shift the spool to the left.

Poppet type solenoid pilot directional control valves

This is a combination valve consisting of a main valve having 4-poppet, solenoid directional control valve for pilot, and pilot selector valve. By varying combination of main valve and pilot selector valve, each poppet can perform directional control, flow control or pressure control.

Manual Directional Control Valves

Directional control of oil flow can also be done by manual operation type of valve shown in Figure 3.12, usually found bundled in series of 5 or more together and are located by the side of a crane lifting trucks or tow trucks.

Control of such valves is done by turning (or shifting) a manual level by the side of the control valve. This shifting action of to and fro results in moving the spool in either direction. They are usually detent type (safety reasons) which means that the ball in the spool is pressed against the spool's groove by spring. Both ways of shifting the spool will return to the neutral position when the lever is released. The load or cylinder position will stay at last position.

S/N	Descriptions	Physical Appearance	Sectional View
1.	Manual Directional Control Valve		
2.	Hydraulic Symbol		

Figure 3.12

Servo Control Valves

Another class of direction control valves are the servo control valves which we shall just discuss briefly. Book II will discuss servo control and proportional valves in more details.

Servo Valve

Hydraulic servo control valves has been used for long time in aircrafts, ships and military installations. However, the demand for high precision applications and needs, more of it application has spread to general industrial purposes.

It covers a wide array of industrial machineries that includes table feeding for machine tools, speed control for main shafts, thickness control for rolling mills, industrial robots, medical devices and various types of testing equipment.

Servo Mechanism

Servo mechanism is an automatic control system which is designed to control mechanical positions by following a given variation in target values. The instantaneous results of the control operation are compared against target values, and a corrective operation is made to agree with the target values all the time. This myriad of corrective actions in the form of numeral values and control operations is known as the **'feedback control'**.

There are many types of servo mechanism according to methods of detecting output displacement, and the types of amplification, transmission and output in the system.

Flow Control Valves

Flow control valves are classified into the following types:

1. Throttling valves

2. Flow Control Valves

3. Deceleration valves

4. Flow control valves with relief valve

5. Flow dividing valves

6. Flow control valves with shut-off valve

7. Pilot Controlled flow control valves

S/N	Descriptions	Physical Appearance	Sectional View
1.	Flow Control Valve		
2.	Hydraulic symbol		

Figure 3.13

Throttling valves

Throttling valves are used to control flow in oil hydraulic circuits. While they have advantages in simple construction, easy operation, and wide control range, flow rate passing through these valves varies with difference in primary and secondary pressures even if they are opened at a constant angle. As a result, they are used where variation in differential pressure is small and high accuracy is not required. Throttling valves with check valve are capable of controlling oil flow in one direction while allowing one in other direction to pass freely.

Flow Control valves (via metering)

Flow control valve has the configuration of pressure compensator (differential pressure – fixed pressure reducing valve) and throttle valve. These combination has negligible effect on the flow rate due to small variation between primary and secondary pressures. Also, they are slightly affected by the change in oil temperature because of sharp edges orifice.

When adjustable flow rate is small, excess flow may occur instantaneously during the operation to jump the actuator. Since this is caused by time delay before the pressure compensated piston reaches its operating position, initial opening of the piston should be adjusted. The adjustment is required when adjustable flow rate is changed or when difference in primary and secondary pressures is varied.

There are two basic types of control situations making use of metering control valves. The following circuits are depicted in Figure 3.14 and figure 3.15.

a. Meter-in control (show circuit diagram)

Meter-in is a method by which a flow control valve is placed in a hydraulic circuit in such a manner that there is a restriction in the amount of fluid flowing to the actuator. Figure 3.14 shows a meter-in control in a hydraulic system.

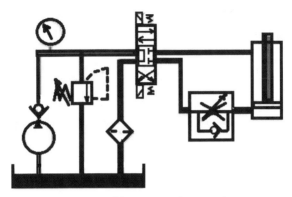

Fig. 3.14 Meter-in Control

If the flow control valve were not to be location, the extension and retraction of the actuator which in this case is a cylinder, would have proceeded at an unrestricted rate. The presence of the flow control valve enables restriction in the fluid flow to the cylinder and thereby slowing down its extension. In the event of the flow direction being reversed, the check valve ensures that the return flow bypasses the flow control valve.

For the same meter-in operation, if it is shifted to the other line. This enables the actuator to extend at an unrestricted rate but conversely the flow to the actuator during the retracting operation can be restricted so that the operation takes place at the a reduced rate. The meter-in control valve is suitable for cases when load during the operation is constantly positive. However, the cylinder would encounter cavitations problem with an overrunning load in which the actuator has no control.

b. Meter-out control

In the meter-out operation shown in Figure 3.15, the direction of the flow through the circuit is simply changed as can be made out from the diagram. It is the opposite to a meter-in operation as this change in direction will cause the fluid leaving the actuator to be metered. The advantage with the meter-out operation is that unlike in the case of meter-in operation, the cylinder here is prevented from overrunning and consequent cavitations.

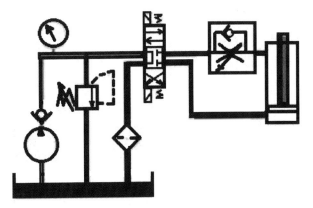

Fig. 3.15 Meter-Out Control

One major problem confronting the meter-out operation is the intensification of pressure in the circuit which can in turn occur on account of a substantial differential area ratio between the piston and the rod. Pressure intensification occurs on the rod side when the meter-out operation is carried out without a load on the cylinder and can result in failure of the rod's seals. It is therefore, seen that both the meter-in and meter-out operations have their relative advantages and disadvantages and only the application determines the type and nature of flow valve placement.
Quite often, you would find this type of valve connected in series to the cylinder exit to limit an amount of oil returning to the tank. Oil supply on the inlet side returns to the tank through the relief

valve. This is used where cylinders might drop faster than speed limit, e.g. vertical drilling machines, and where back pressure needs to be applied to cylinders all the time. Care should be taken for conditions on cylinder rod side which may cause the pressure higher than relief pressure.

Shuttle valves

This is another type of direction control valve. It allows a system to operate from either of two fluid power sources. One application is for safety in the event that the main pump can no longer provide the hydraulic power to operate emergency devices. As soon as the primary source is exhausted, the shuttle valve shifts to allow fluid to flow from the secondary backup pump. This type of valve is shown in Figure 3.16.

S/N	Descriptions	Physical Appearance	Sectional View
1.	Shuttle Valve		
2.	Hydraulic Symbol		

Figure 3.16

The shuttle valve consists of a floating piston, which can be shuttled to either side of the valve depending on which side of the piston has a greater pressure. Shuttle valves may be spring loaded in one direction to favour one of the supply sources. The shuttle valve is essentially a direct acting double check valve with a cross bleed, as depicted by the symbol. The double arrows in the symbol indicate that reverse flow is permitted.

Bleed off valve (show circuit diagram)

Following Figure 3.17 shows a bleed off valve, typically it is also a needle valve. It is basically a flow control valve but without a check to it direction. The amount of oil flow can be varied by reducing the orifice of the valve. It can be totally shut off.

S/N	Descriptions	Physical Appearance	Sectional View
1.	Bleed Valve		
2.	Hydraulic Symbol		

Figure 3.17

Flow control valves are installed in a bypass line from a main circuit to control an amount of oil returning to a tank and thus vary the speed of actuators. Since pressure in the main circuit equals to load resistance, heat generated by surplus oil is less than meter-in or meter-out control. However, variation in discharge from the pump also directly affects the speed of actuator because a constant amount of oil flows into the actuator cannot respond to load variation situations. Therefore, the change in pump's volumetric efficiency and accuracy required for actuator speed should be considered before designing the circuit. This cannot be deployed when two or more actuators are operated at the same time.

Deceleration Valves

This valve is used for purpose of slowing the cylinder speed as it restricts the flow of hydraulic oil continuously with a cam mechanism. In the normally open type, flow is restricted by pushing the spool down. For some other designs, like the normally close type, flow is increased by pushing the spool down. When the normally open type is used as cushion at an end of the cylinder stroke, it is difficult to determine cylinder's stop position precisely by cam. In this case, the cylinder may be slowly moved to a specified position by controlling the throttling valve and then stopped by the directional control valve.

Flow control valves with relief valve

Flow control valves of this type are designed to compensate for a disadvantage of meter-in control in that situation whereby the pump's discharge pressure rises to a level set for oil hydraulic circuits even if the load pressure is low. Combined with special relief valve, the discharge pressure is controlled to the load pressure plus 6 kgf/cm^2 (85 psi)

Flow Divider Valves

Flow divider valves are used to synchronize operations of two or more actuators. They have two outlet ports (A and B) with a common inlet port (P) as in Figure 3.18. They are classified into flow dividing valves which provide synchronized control for operations of actuators in one direction, and flow combining valves which provide synchronized control in both directions. However, the intended synchronized cylinders must be of similar sizes and stroke lengths.

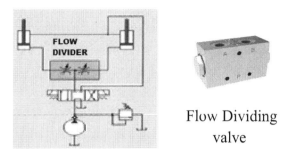

Flow Dividing valve

Figure 3.18

Stackable Valves System (CETOP/NFPA)

Stackable valves also known as sandwich valves are commonly used in many installations for hydraulic power system. Their physical features for many made are misleading without examining them closely for their functions. It is usually marked or displayed with a type plate on the valve body itself.

Size of all the valves are however, fixed with unified standards. For examples, the valves standard sizes are descript as CETOP 3, 5,7,8 and 10 or NFPA D01, D02, D06 & D10. Manufacturers made according to this standard for purpose of interchange ability and replaceable. These 'CETOP' valves are lined or stacked for the following reasons.

(NB: Figure 2.7 (3) shows a HPU with neatly stacked control valves to the left side of the electric motor).

1. Save space, ease of installation and reduce repeated pipe works drastically.

2. They can be assembled easily, together with ease and any future modification to the circuit can be executed easily.
3. Since no piping is required between the valves, oil leak, vibration and noise problems are minimized.

4. The valves stacking also reduce the tank size and overall weight therefore; maintenance and inspection can be done efficiently.

On/Off Valves

There are several types of commercially available on/off valves. Those commonly used to handle high pressure fluid control valves are shown in Figure 3.19.

S/N	Descriptions	Physical Appearance
1.	Ball Valves	
2.	Needle Valve	

Figure. 3.19

On and off valves, sometimes also known as shut-off valves are common valves used to handle high pressure in hydraulic system. They can be simply ball valves or needle valves (in the form of plug and piston designs). For ball valves, you have to take note that it is for high pressure applications within the intended pressure and service media. The specifications and applications are marked clearly on the valves' body denoted by WOG, meaning for water, oil or gas services, with service pressure rating in PN (DIN standard) or PSI (ANSI standard) and size..

The standard sizes of these valves comprise of 1/8", ¼", 3/8", ½", ¾", 1" and many more others. It also come in many forms of port-connections standards. The most common ports connections are finished with standard threads available in BSP and NPT. Please check with the manufacturers and suppliers for details.

Manifolds

Leaky fittings are a cause for concern in hydraulic circuits especially with increased in the number of connections and piping works. At time, it can get very messy, especially without proper design and planning. This is where manifolds play a very important role.

Their incorporation in a hydraulic circuit helps drastically to reduce the number of external connections required. Figure 3.20 shows a simple manifold commonly used in hydraulic systems.

Figure 3.20

In the case of modular valve stacking, the manifolds used are provided with common pressure **'P'** and return ports **'T'**, with each valve station being incorporated with individual A and B work ports. These four ports 'P, T, A and B' are clearly marked or indented on the manifold body.

Manifolds are normally specified according to system pressure, total flow, number of work stations, valve sizes as according to CETOP or NFPA porting and pattern.

Chapter 4 – Actuator

Actuators
Hydraulic cylinders
 - Tie-rod & welded end design
Type of Hydraulic Cylinders
Single Acting Cylinder
Double Acting Cylinder
Telescopic Cylinder
Hydraulic Jack
Hydraulic Motors
 - vane, gear, piston & oscillating types

Upon completion of this chapter, you should be able to:

1. Know about most hydraulic actuators
2. The various types of actuators and their specific uses and applications
3. Know about each type of actuator, it's capabilities, capacities and functions
4. Recognise the type of actuators used on site

If the rate of change on the outside exceeds the rate of change on the inside, the end is near.
Jack Welch

Actuators

Actuators are devices to convert hydraulic energy of oil to mechanical energy and are classified into the following types:

Hydraulic cylinders
Hydraulic motors

Oscillating motors.

Hydraulic Cylinders

Hydraulic cylinders come in many types and versions. Two major classifications must be noted, mainly Tie-rod cylinder type and the welded end cylinder as in Figure 4.1 (1 & 2). below.

S/N	Descriptions	Physical Appearance
1.0	Hydraulic Cylinder Type 1 (Tie-Rod)	
2.0	Hydraulic Cylinder Type 2 (Welded-End)	
3.0	Telescopic Cylinder	
4.0	Hydraulic Jack	

Figure 4.1

In term of hydraulic cylinder physical features and designs, they are quite diverse as manufacturers have their own production programs. However, there are standards bodies with guidelines and specifications for hydraulic cylinder constructions like NFPA or other equivalent standards. Basically, hydraulic cylinders consist of cylinder honed tube, hard-chromed piston rod and covers (known as head and end caps), cushioning device and constructions for mounting purposes.
(NB: NFPA stands for National Fluid Power Association)

Cylinder Hone Tube

The cylinder barrel is made of a seamless thick-wall cylindrical tube or pipe that must be machined internally and finished it by honing process (similar to a grinding process). It has wide ranges of standard sizes by imperial inches and metric. The internal diameter of the cylinder or it technical term 'bore size' in inches range from 1 ½" to 18". The bore sizes of the cylinder in metric measurement are available from 40mm to 250mm. The cylinders with bore size above the standard sizes are considered as non-standard.

Cylinder Base or Cap and Cylinder head piece

In most hydraulic cylinders, the barrel and the bottom portion are welded together. The heat from welding process can damage the internal honed surface of the barrel if done poorly. Before this welded end design was available, older version of hydraulic cylinder designs have a screwed or flanged connection parts consisting of the base cap and cylinder head piece assembled together by 4 pieces of tie rods with threaded ends. More tie rods are required and usually found on larger ones and those cylinders to handle high pressure. In some designs, the cylinder head is sometimes connected to the barrel with a sort of simple lock. The other threaded ends are usually tightened and under tension by sets of nuts and lock nuts.

However, flange connections are the best but much more expensive. This is because the flange has to be welded to the cylinder before machining. This method of assembling has several advantages over the rest of the tie rod designs because of ease of parts standardization, speed up assembling and ease of services and repairs. For larger bore size cylinders of diameter 300mm and above, it poses a challenging problem on handling as well as alignment during assembling. Therefore, the larger the bore sizes, only a handful of manufacturers are competent or have the machines facilities to produce it due to it huge sizes.

Piston

The piston is a short cylindrical metal component that separates the two parts of the cylinder barrel internally. The piston is usually machined with an internal cylindrical hole to receive one end of the piston rod (describe below) and it outer surface with grooves to fit 'U' of 'C' cub seals with lips, split fibre ring and sometimes metal seals and others. They prevent the pressurized hydraulic oil from passing through the piston to the chamber on the opposite side. This difference in pressure between the two sides of the piston causes the cylinder to extend and retract depending on oil flow directions through the two ports on the two ends of the cylinder barrel.

Piston seals vary in standard sizes, design and material according to the pressure and temperature requirements that the cylinder is being designed for a particular service. Generally, most hydraulic seals kits are made of nitrile rubber, viton (to handle higher temperature) or other materials depending on the type of services. There are also some that made of composite materials for very extreme purposes.

Piston Rod

The piston rod is a hard chrome-plated piece of cold-rolled steel which attaches to the piston and extends from the cylinder through the rod-end head. The other end of the piston rod usually machined with a collar that rest on the piston and a threaded portion protruded through the piston and held together with a nut (sometimes machined) fully restraint from movement by mechanical means. The piston rod connects the hydraulic actuator to the machine component intended to do the prescribed work. There are several connections and mounting system in the form of male or female clevis attached to the piston rod end. Most manufacturers will offer the standardize mountings for competitive reasons while some may even offer customized features with a small fee. However, in the competitive world, most manufacturers are prepared to waive this small fee if you have quantity to justify their production costs.

Rod Gland

The cylinder head is fitted with seals to prevent the pressurized oil from leaking past the interface between the rod and the Cap head piece. This area is called the rod gland, it often has another seal called a rod wiper which prevents contaminants from entering the cylinder when the extended rod retracts back into the cylinder. The rod gland also has rod wear ring. This wear ring acts as a linear bearing to support the weight of the piston rod and guides it as it passes back and forth through the rod gland. In some cases especially in smaller bore sizes hydraulic cylinders, the rod gland and the rod wear ring formed a single integral machined part. In this case, usually it is assembled with the barrel by threaded means, sealed with set of seals kit and o-rings and locked in place with an internal cir-clip.

Tie Rod

The tie rod type hydraulic cylinders use high strength threaded steel rods to hold the two ends caps to the cylinder barrel. This method of construction is most often seen in industrial factory applications. Small bore cylinders usually have 4 tie rods, while large bore cylinders may require as many as 16 to 20 tie rods depending on bore sizes of the cylinder. This is because the hydraulic forces of such cylinder sizes can be of tremendously high during operations and therefore it requires the numbers of tie rods to meet the adequate strength.

NB: The National Fluid Power Association (NFPA) has standardized the dimensions of hydraulic tie rod cylinders. This enables cylinders from different manufacturers to interchange with their mountings and fittings.

Welded body cylinder

The welded body cylinder is very similar to the tie-rod design. This cylinder, you will find that the end cap is welded to the hone tube (the main cylinder body), thus doing away the tie-rods. The rest of the major components and internal parts are very similar to the tie rod construction except that it is without any tie rod. Figure 4.1 (2.0) shows the welded end cylinder.

Cushioning Device

The cushion device is designed to reduce shock generated by piston hitting the cover at its stroke end during high speed reciprocating action in the cylinder, thus prolonging the cylinder's life and eliminating influence of impact wave (term as back pressure or unwanted shock wave propagating downstream) on hydraulic circuits and auxiliary equipment.

For the cushioning device, particular care should be taken for increase in back pressure when the piston reaches the cushion stroke. This does not cause a serious problem on the head side, but it can be damaging on the piston rod side. This is because if the cylinder is subjected to downward load, the back pressure becomes a sum of internal pressure balancing with load and the pressure increased by the ratio of areas on the rod side to that on the head side. The resultant pressure can sometime be in excess of the allowable stress for cylinder tubes. It could lead to material deformation or destruction over time and also result in premature failure. Therefore, care should be exercised for piston speed, especially when the piston approaching the end of stroke. If it is excessively fast, even with a fully open adjuster, it will not help.

Ports Opening

All hydraulic cylinders consist of 2-port openings, usually have the same size. Each port opening is located at both ends (lower and upper ports) on the side of the hone tube. You may also find some customised cylinders with lower port located at the end cap. For tie rods design cylinder, it is commonly found on the cap head and end cap piece.

The sizes of the port openings for both types of designs are made according to standard threads of BSP or NPT. All manufacturers usually have one of these thread standards in their production program but they could provide any one of these or other known threads standard on special request. The porting sizes can range from 3/8" to 1 1/2" depending on the bore sizes of the cylinders.

Type of Hydraulic Cylinders

Hydraulic Cylinders can be constructed in the following designs.
1. Single acting
2. Double acting
3. Telescopic types
4. Hydraulic Jack

Single Acting Cylinders

This type of cylinder usually enclosed with an internal compression spring. When power up, the oil enters through one port (usually at the lower port) to push the cylinder rod outward. The oil pressure have to overcome the spring force and push the desire load. The returning stroke will be depressed by the load and the enclosed spring force.

The advantages of using single acting cylinder are that it only requires a single hose connection and usually only require a simple 3-way control valve on the control end.

Double Acting Cylinders

This type of cylinder as shown in figure 4.1 consists of 2-port openings. When power up, the oil enters through the lower port of the cylinder to push the cylinder rod outward. For the return stroke, the oil now enters through the upper port on the cylinder. Oil pressure push the piston together with the rod downward. The retracting piston pushes the oil out through the lower port.

For double acting cylinder installation, it requires two hoses (or piping) connection in the hydraulic system. The two other ends are usually connected to A and B ports of a 4-way directional control valve on the control end.

The main advantages are the delivery of various output forces and the cylinder speed can be varied as desired.

Telescopic Cylinders

For some applications, the standard single acting and double acting cylinders cannot meet all the requirements especially due to stroke length limitation, telescopic cylinder is the solution to many of the needs. You can find these in high tonnage lifting cranes or some hoisting trucks where they have multi-stages cylinders.

Basically, telescopic cylinders consist of more than 2 stages and are usually double acting design. This is because a single acting telescopic cylinder will not function properly as it requires an acceptable external load to retract the cylinder. It is not feasible in most practical situations. Therefore, you will find more of double acting telescopic cylinders than the single acting type in most installations. A double acting telescopic cylinder can be extended and retracted hydraulically as the normal double acting cylinder.

However, there is also limitation with this telescopic cylinders. It should never be deployed as direct structural members for machine elements or equipment. It is not rigid enough to provide that stable structural support. Nevertheless, it is an excellent device to generate force over extremely lengthy stroke-length which normal double acting cylinders cannot achieve.

Hydraulic Jacks

Hydraulic jacks can be in the form of single acting (single port, usually lower port, oil entry and exit) or double acting (2-port) construction. The differences between normal cylinder and jack are that the latter has thicker walled cylinder than the hone-tube used by the normal cylinder and larger rod size to handle high pressure above 10,000 psi (700 bar). It is usually extended with a hydraulic hand pump. Their design and construction features are designed for extremely high tonnage capacity load.

It is commonly found in buildings and ship building and constructions facilities. The most common use of these jacks are for load test on piles' foundation for buildings construction.

Hydraulic Motors

Hydraulic motors are designed to convert fluid power to mechanical power of rotation. In hydraulic motors, the rotating speed can be varied easily by controlling the supply flow rate of control valves.

They are classified into vane type, gear type, and piston type with high speed use and low speed high torque use. The constructions are quite similar to hydraulic pumps and must not be confused.

Vane Motors

Vane motors have similar construction to that of pressure balancing vane pumps. However, since the vane cam ring need to be in contact at the starting time, the spring is used to push up the vane.

The internal construction of the vane motors is similar to that of a vane pump; however the principle of operation differs.

Vane motors develop torque by virtue of the hydraulic pressure acting on the exposed surfaces of the vanes, which slide in and out of the rotor connected to the drive shaft. As the rotor revolves, the vanes follow the surface of the cam ring because springs are used to force the vanes radially outward.

No centrifugal force exists until the rotor starts to revolve. Therefore, the vanes must have some means other than the centrifugal force to hold them against the cam ring. Some designs make use of springs, while other types use pressure-loaded vanes. The sliding action of the vanes forms sealed chambers, which carry fluid from the inlet to the outlet.

Gear Motors

Like gear pumps, gear motors are also classified into internal gear and external gear types. Unlike gear pumps, however, gear motors rotate in both directions and have an external drain because of pressure working on both ports to ensure constant suction and discharge when in operation.

Gear motors are simple in construction. A gear motor develops torque due to the hydraulic pressure acting on the surface of the gear teeth.

By changing the direction of the flow of fluid through the motor, the direction of rotation of the motor can be reversed. As in the case of a gear pump, the volumetric displacement of the motor is fixed. The gear motor is not balanced with respect to the pressure loads. The high pressure at the inlet, coupled with the low pressure at the outlet, produces a large side load on the shaft and bearings, thereby limiting the bearing life of the motor.

Gear motors normally limited to operating pressures of around 140 Kg/cm^2 (about 2,000 psi) and operating speeds of 2,400 rpm. They are available with maximum flow capacity of 550 lpm (145 gpm).

Hydraulic motors can also be of the internal gear type, like the gear pump. The internal gear type motors can operate at higher speeds and pressures. They also have greater displacements than the external motors.

The main advantages associated with gear motors are its simple design and cost effectiveness. They also possess good tolerance to dirt. The main disadvantages with gear motors are their low efficiency levels and comparatively higher leakages.

Piston Motors

Piston motors work under the principle reverse to that of piston pumps; oil is introduced to the piston units of cylinder to convert fluid energy to mechanical energy which then causes the rotation of the piston motor drive shaft.

Piston motors are classified according to arrangement of pistons and output shafts, as follows;

1. Axial motors (pistons are arranged parallel to shafts) – Bent axis type and Swash plate type.

2. Radial motors (pistons are arranged perpendicular to shafts) – Eccentric type and multi-stroke type.

Oscillating motors

In contrast to hydraulic motors which perform continuous rotary actions and hydraulic cylinders which perform reciprocal linear actions, oscillating motor perform oscillating rotary actions. There are the single vane type that oscillate at 280° and the double vane type that oscillate at 100°. Oscillating motors have the following advantages over other conventional mechanism.

1. It is better than conventional form of linkages, like design faces space constraints.

2. Higher torque over other cumbersome linkages can be achieved.

S/N	Descriptions	Physical Appearance
1.	**Gear Motor**	
2.	**Vane Motor**	
3.	**Piston Motor**	

Figure 4.2

3. High efficiency in term of mechanical advantage as oil hydraulic is transmitted directly.

4. The construction is simple and compact thus it is easier to handle than rack pinion system and piston or helical spline system.

Points to Note about hydraulic pumps and motors:

These hydraulic motors as in figure 4.2 appear to be similar to their respective pump types. For example, the piston pump and piston motor are almost identical twin in their physical appearances without examining their respective type-plate pinned onto their casings.

However, if you examine both closely, you would notice something different physically, of course besides where they are being located in the installation. The pump's inlet or suction port size is larger than the outlet port size. Whereas the motor has practically equal inlet and outlet ports and they are bi-directional.

Chapter 5 – Hydraulic Basic Accessories

Oil Filters and Tank Filters
Oil heater/Cooler
Measuring Instruments
Accumulators and applications
 - **Bladder**
 - **Diaphragm**
 - **Piston**
Hydraulic Oil
Selection Criteria of Hydraulic Fluid
Viscosity of Hydraulic Fluid
Tank size and Tank Accessories
Electric Motor and selection
Steel Piping and Fluid Power Transmission
Pipe Schedules and Codes
Steel Tubing and Flexible Hoses

Upon completion of this chapter, you should be able to:

1. Know about most filters and its function
2. The various types of accumulators and applications
3. Know about various important components in a hydraulic tank
4. Recognise the types of fittings, pipe's schedules and hydraulic hoses
5. Know the basic steps and requirements for sizing a hydraulic system

The easier it is to do something, the harder it is to change the way you do it.
Steve Woznaik

Oil Filters and Tank Filters

Hydraulic filters are used to separate foreign matters from mixing with hydraulic oil and are classified into line filters used in pressurized lines and tank filters installed on the pump suction side. By making best use of these filters, efficiency of hydraulic equipment can be improved together with longer service life and better economy.

Tank filters come with or without case. Tank filters without case are generally used inside the tank storing hydraulic oil. On the other hand, tank filters with case are installed outside the tank and are equipped with indicator to show clogging of filter elements.

Tank filters without case

Tank filters are made of light alloy fine mesh core around to form a cylindrical tube. The aluminium is specially treated and some of the designs come with stainless wires wrapped around for reinforcement. Their filtration rating is usually 150 mesh size. The filtration area in most designs will indicate a recommended rated flow and have to meet or maintain well below 50 mmHG. There are also some other tank filters uses bronze metal mesh.

It is mounted on the suction side of pump via a short threaded pipe in which the pipe's size is consistent to the size of the port size of pump.

Tank filters with case

These types of filters are installed outside the tank and consist of a filter element housed with a cylindrical outer casing which can be inspected easily. Some of the designs are equipped with indicator and limit switch to allow the operator to check for filter clogging.

Pipe line Filter

Tank filters cannot have very small filtration rating because of suction resistance. Pipe line filters are used to cover the disadvantage of tank filter.

Pipe line filters are used to remove dust and dirt contained in hydraulic oil circulating through pipes, and is generally classified according to pressure stage and filtration rating.

Filter elements used in pipe line filters include notch wire type, wire net type, sintered metal type, and paper type. The most suitable type should be selected in consideration of types of equipment, from general hydraulic equipment to servo hydraulic equipment.

Particularly, servo hydraulic equipment uses mainly stainless wire net elements because high filtration rating is required on the secondary side. Moreover, servo control hydraulic systems are highly precision equipment which is sensitive to dirt or particle.

Some filters have magnet or stop valves built in to facilitate filter element change without much interruption to the whole system.

Oil heater/Cooler

Oil coolers are essential in removing heat generated in hydraulic circuit, thereby maintaining temperature of hydraulic oil at an appropriate level and ensuring smooth and efficient operations. Generally, the water cooling type is mostly used. Also, the air cooling type is used where use of water is not desirable or cooling water is not readily available.

Measuring Instruments

Pressure gauges are the most common piece of measuring instrument to measure the pressure of the hydraulic pressure line. It is installed in conjunction with the pressure setting valve (PRV) to allow indication of the desired pressure for the system. It includes Bourdon tube type pressure gauge and strain gauge type pressure detector. The most appropriate type should be selected in consideration of performance and operating conditions.

Pressure Gauges are calibrated according to standards for their accuracy and come in the form of bottom mount or back mount with BSP and NPT threaded connections. The Pressure range for hydraulic applications is typically 0 - 250 bar (255 kg/cm^2) for metric unit and 0 - 3,625 psi for imperial uiit. They are available in 63mm (2") and 100mm (4") in diameter, usually stainless steel casing with clear glass and options of glycerine filled or none. High pressure gauges of between 600 to 700 bar (8,700 - 10,150 psi) are also available for high pressure system.

The reading from a gauge is approximately termed as *gauge pressure*. Absolute pressure, on the other hand, would read zero only in a complete vacuum. Thus, gauge and absolute pressures are related by,

Absolute pressure = Gauge pressure + Atmospheric pressure

NB: Absolute pressure ((bara or psia) and Gauge Pressure (barg or psig).

Accumulators and Applications

Accumulators are devices, which simply store a potential energy in the form of hydro-pneumatic under pressure. It incorporates a gas (usually nitrogen gas) in conjunction with a hydraulic fluid. The hydraulic fluid has minimal power storage qualities, but the gas can be compressed to high pressures and low volume. Potential energy is stored in this compressed gas to be released upon demand.

Figure 5.1 shows the 3 most popular accumulators widely used in various industries.

S/N	Descriptions	Physical Appearance	Sectional View
1.0	Bladder Type		BLADDER
2.0	Diaphragm Type		DIAPHRAGM
3.0	Piston Type		PISTON

Figure 5.1

In all types of accumulators, for example of a piston type, the energy is stored in the compressed gas exerts pressure against the piston separating the gas and hydraulic fluid. The piston in turn forces the fluid from the cylinder into the system and to the location where useful work will be accomplished.

There are three main types of accumulators available commercially, namely, the weight type, bladder type and the piston type. Of the three, bladder and piston are the two most commonly used in all industries including the ship and building services and marine.

Weight-Loaded Accumulators

The weight-loaded type is the oldest type of accumulator. It consists of a vertical heavy wall steel cylinder, which has a piston with seal (or packing) to prevent leakage. It is less popular but able to discharge stored pressure at a constant level. The only disadvantage is it large size.

Bladder Accumulator

The bladder accumulator contains an elastic barrier between the oil and gas as shown in the Figure 5.1 (1). The bladder is fitted to the accumulator by means of a vulcanized gas-valve element that can be installed or removed through the shell opening at the poppet valve. The poppet valve closes the inlet when the bladder is fully expanded. This prevents the bladder from being pressed into the opening. A shock-absorbing device, protect the valve against accidental shocks, during a quick opening.

The greatest advantage with these accumulators is the positive sealing between the gas and oil chambers. The lightweight bladder provides a quick pressure response for pressure regulation as well as applications involving pump pulsations and shock dampening. As such this bladder type accumulators are most widely used.

However, it becomes instantly inoperative once the bladder is damaged. Selection criteria should be based on uses and purposes.

Diaphragm Accumulator

The diaphragm type accumulator consists of a diaphragm secured in a shell and serving as an elastic barrier between the oil and the gas. The cross section of a diaphragm type accumulator is shown in Figure 5.1.

A shut off button which secured at the base of the diaphragm, covers the inlet of the line connection when the diaphragm is fully stretched. This prevent the diaphragm from being pressed into the opening during the pre-charge period. On the gas side, the screw plug allows control of the charge pressure and the charging of the accumulator by mean of a charging and testing device.

Piston Accumulator

This accumulator consists of a cylinder containing a freely floating piston with proper seals. The piston serves as a barrier between the gas and oil. A threaded lock ring provides a safety feature that prevents the operator from disassembling the unit while it is pre-charged.

The main disadvantage of piston type accumulators is that they are very expensive and have size limitation. In low-pressure systems, the piston and seal friction also poses problems. It is not advisable to use piston accumulators as pressure pulsation dampeners or shock absorbers because of the inertia of the piston and the friction in the seals.

The principle advantage of the piston type accumulator lies in its ability to handle very high- or low-temperature system fluids, through the utilization of compatible O-ring seals. The piston type is slower in response than the bladder type, but it is more reliable because it is less vulnerable against damage.

Generally, there are several advantages for using accumulators in the hydraulic design and system. They are to serve the following purposes, besides storing of energy.

1. Absorbs Pulsations

In most fluid power applications, pumps are used to generate the required power to be used or stored in a hydraulic system. Many pumps deliver this power in a pulsating flow. The piston pump is one example, as commonly used for higher pressures application and it tends to produce tremendous high pressure.

In this case, it is a good practice to use an accumulator to absorbs pulsation and if properly located in the system or circuit, it will substantially cushion these pressure variations.

2. Cushions Operating Shock

In many fluid power applications the driven member of the hydraulic system stops suddenly, creating a pressure wave (back pressure), propagates backward through the system. This shock wave can develop peak pressure several times greater than normal working pressure and can be the source of system failure. The gas cushion in an accumulator, properly placed in the system will check or minimize this shock.

3. Supplements Pump Delivery

An accumulator is capable of storing power and can supplement the fluid pump in delivering power to the system. The pump stores potential energy in the accumulator during idle periods of the work cycle. The accumulator transfers this reserve power back to the system when the cycle requires emergency or peak power. This enables a system to utilize a much smaller pump, resulting in savings in cost and power.

4. Maintains Pressure

Pressure changes occur in a hydraulic system when the liquid is subjected to rising or falling temperatures. There may also be a pressure drop due to leakage of hydraulic fluid. An accumulator compensates for such pressure changes by delivering or receiving a small amount of hydraulic liquid. In the event of power failure or accidentally cut-off, the accumulator would act as an auxiliary power source, maintaining power in the system.

Hydraulic Oil

Table below indicate the basic classification of hydraulic oil

Table 5.1 - Hydraulic Fluids

S/N	Petroleum Fluid	Hydraulic Oil	
		Flame Resistive Hydraulic Fluid	
		Synthetic Fluid	Water base type hydraulic fluid
1.	Additive Turbine Oil	Fatty acid esters fluid	Water/glycol hydraulic fluid
2.	Special Fluid	Phosphate ester fluid	Water-in-oil type Emulsion/(w/o emulsion)
3.	General hydraulic Fluid a. High viscosity index fluid b. Low temperature fluid c. High temperature fluid d. Abrasion resistance working fluid		Oil-in-water type emulsion
	NB: Biodegradable hydraulic oil is another class of oil gaining popular as it is environmentally friendly. These fluids are generally based on naturally vegetable oils and are biodegradable by naturally occurring organisms. However, large quantities of spills will still need to be handled like the other mineral oil spills. **Another point to note is, always check specifications before use.**		

Selection criteria of hydraulic fluid

Hydraulic equipment is one type of fluid machine requiring hydraulic oil for operation. It is very common that hydraulic pumps and valves are operated at high pressure and high speed. As a result waste heat generated convey through oil in contact with mechanical moving parts place a lot of stress on the whole system. In addition, there are also other factors like the correct temperature for operation, materials selection for the equipment and work place environmental factors. Hence, selecting the correct hydraulic oil for the working fluid is very important.

The following are important points to consider when selecting the working fluid for the hydraulic system.

1. Always ensure the right viscosity if system need to work with wide temperature band.

2. Oil will still maintain fluidity at low temperature and not deteriorate easily when used under high temperature.

3. Must have good oxidation stability.

4. Must have good shearing stability.

5. Must have rust proofing capability.

6. Have good anti-foaming tendency.

7. Must be fire resistance.

Points to note:

Viscosity of Hydraulic Fluid

Viscosity of hydraulic fluid is expressed by dynamic viscosity (m^2/s), which we had discussed in chapter 1. The correct viscosity of hydraulic oil for a piece of hydraulic equipment is almost like 2nd nature!. Inappropriate viscosity will result in poor suction, excessive heat generation, internal leak, poor lubrication, malfunction of valves in the whole system. The equipment would encounter shorter life span or even result in serious accident.

A general guide for most hydraulic applications, for light duty use hydraulic number 32. For medium duty use Hydraulic number 46. Table 5.2 is a general guide may be consulted for specific pump used in your hydraulic system.

Table 5.2 Appropriate viscosities according to type of pumps

Viscosity (cSt)	Pump Types			
	Vane	Gear	Piston	
			Axial	Radial
Minimum Viscosity	20	16 ~ 25	12	16
Appropriate viscosity	25	25 ~ 70	20	30
Maximum Viscosity	400 ~ 800	850	200	500
Suction capacity (-mmHg)	250	400	25	100

Tank Size and Accessories

In all hydraulic installations whether they are small or large, the hydraulic power unit will always start with the tank and its accessories mounted on top of it. These include, pump and electric motor set (or some other prime movers), relief valves, directional control valves and those auxiliary equipment mentioned earlier.

The basic requirements for tank construction are as follows:

1. The tank usually designed with an enclosure must be able to prevent dust and other foreign matters from entering. Air vent or breather, like the petrol-powered car refilling cap is incorporated usually on top or at a corner where it is double use as oil refilling.

2. The tank unit should be made to be simply detachable from the main unit in order to facilitate the maintenance and to ensure the system working accuracy after re-connection.

3. The purpose of the air vent is there as a check for vacuum effect during heavy demand of oil supply while also preventing excess pressure drop on the pump delivery system.

4. The capacity and size of the tank is the most critical requirement for attention as insufficient oil reservoir in the tank due to physical size will bring about disruption for intended operation cycle. Having said this, the designer must ensure oil level at appropriate level at all time in order to cater for smooth operation. A conservative designer would also consider future extension or minor alteration to the processes requested by end user.

5. For larger tank servicing varieties of actuators, it is wise to include a baffle plate between suction pipe and the return pipe to separate sediment from foreign matters. There is one other advantage, that is to check air bubbles from the return line which can flow quickly into the suction pipe creating cavitation.

6. The oil level gauge or known as the sight glass apparatus is a compulsory item to indicate a safe oil level limit.

7. The rule of thumb for sizing a tank reservoir will depend on following parameters

a. The type of pump use and delivery.

b. The actuator capacity.

c. Distance between the tank (power unit) and rhe sources (actuators).

d. Frequency of use or duty (whether it is occasionally, intermittent or continuous cycle).

For most cases in normal application, we can simply give an estimated tank size based on the following parameters,

1. The size of the pump:

Allow approximately 3 to 6 times of the flow capacity of pump.

2. The size of Actuator(s) in the whole system:

Allow approximately 3 times or more depending on duty (once in a long while, intermittent or continuous). Note also the distant of various actuators from tank installation.

NB: An example is illustrated in chapter 7 showing the most practical steps to size the various hydraulic components.

Electric Motor and Selection

Electric Motors

Hydraulic pumps as we have covered in chapter 2, are pumps which circulate hydraulic oil, the working fluid, through a hydraulic system. They are almost always driven by electric motors. All electric motor consist of a number of coils wound on to a soft iron rotor or armature. The commutator is divided into a number of sections, two for every coil on the rotor. A rotor shaft runs through the centre of the motor and it is this shaft that rotates as the coils rotate. This is the shaft that connects directly to the pumps' shaft with a coupling.

Electric motors can be broadly classified by the type of power source needed to operate them. It has been called the "work horse" of industry. The three major categories of motors are the dc motor, the ac motor and the universal motor. The universal motor is designed to operate from either an ac or a dc power source (refer to next chapter). Of the 3 types, dc and ac motors are the most common. They are being used as the main driving system for almost everything. They also powered almost all hydraulic systems.

Some motors can be further subdivided. For example, there are several kinds of dc motors, brush motor, brushless motors, stepper motors, and servo motors. As for the ac motors there are induction types and synchronous types. For hydraulic applications, usually the induction types with single phase as well as numerous kinds of poly-phase (especially 3-phase) ac motors are widely used.

The general formula to calculate the speed of ac motor is as follows:

$$S = \frac{120 f}{P_n} \quad \text{..................(Eq. 5.1)}$$

S = synchronous speed in revolutions per minute

f = frequency in hertz Hz, usually 50 or 60, where $f = \frac{1}{T}$

P_n = number of poles

Example 5.1

To determine the synchronous speed of a 4 pole motor operating from a 220V, 50 Hz source,

$$\textbf{Synchronous speed} = \frac{120 \times 50}{4}$$

$$= 1,500 \text{ rpm}$$

(NB: If the source is 60 Hz, the synchronous speed = 1,800 rpm)

Points to note:

1. NEMA has developed standards for the dimensions of the various sizes of motor enclosures. The size (height, length, shaft diameter, etc) is indicated by a frame number.

2. Depending on the style of enclosure, motors can be classified as either open motors or totally enclosed motors.

3. Drip proof motor is an open motor in which the openings are

designed so that particles or drops striking the motor enclosure at an angle no greater than 15° from the vertical will not impair the operation of the motor.

4. Splash proof motors are designed so that matter splashing within specified angles will not harm the motor.

5. Totally enclosed fan-cooled (TEFC) motors have an external fan, attached to the rotor shaft, which forces air circulation around the motor.

6. Explosion proof motor is totally enclosed and designed so that an explosion of a gas inside the motor will not leak out to cause another explosion. Usually this class of motors are found in services for oil and gas industries or sensitive installations.

7. Motor rating are parameters for which electric motors are commonly rated include voltage, current, power, speed, temperature, frequency, torque, duty cycle, service factor and efficiency.

8. Service factor of motor is another selection criterion, in some applications, a motor is designed to operate only intermittently. As there are also motors built for continuous duties.

Steel Piping and Fluid Power Transmission

Efficient transmission system of power from one location to the other is a key element in the design and performance of a hydraulic system. this is known as fluid conducting. Fluid conductors comprise that part of the hydraulic system that is used to carry fluid to the various components. These conductors include the likes of steel tubing, steel pipes and hydraulic hoses.

In an actual hydraulic system, the fluid flows through a distribution network consisting of pipes and which carry the fluid from the reservoir through the operating components and back to the reservoir. Since power is transmitted throughout the system by means of this network. It is highly imperative that this network be properly designed to ensure efficient operation.

Steel Piping

Steel pipes are often preferred over other conductors from the standard point of performance and cost. However, since welding is required to be carried out on these pipes to ensure maximum leak protection, they are quite difficult to assemble. Another factor to be concerned with is that they require costly flushing during start-up, to ensure a contaminant free environment. Although steel pipes are specified by their nominal outside diameter, their actual flow capacity is determined by their internal area (or ID).

Pipe Schedules and Codes

In the past, all piping was designated as standard, extra strong and double extra strong. That system allowed no variation for the wall thickness. Also, as the piping requirements started increasing, a greater variation was needed for specifying the pipes. As a result, piping today is classified according to a schedule. The most common schedule numbers in practice are 40, 80, 120 and 160. For pipe diameters ranging from 1/8" (15mm) to 10" (250mm), the dimensions of standard steel pipe correspond to schedule 40.

The standard steel pipes correspond to schedule 40 pipes with a wall thickness of 6 mm. The dimensions of extra strong steel pipes are the same as schedule 80 pipes ranging from diameter 15 to 200 mm.

Due to the increasing variety and complexity of requirements for piping, a number of engineering societies and standards (like ASTM, American Society of Testing, API, American Petroleum Institute and ANSI, American National Standards Institute) have devised codes, standards and specifications that cater to a vast majority of applications.

The pipe schedules commonly used in hydraulic systems are 40, 80 and 160. The classification has been done based on their nominal size and schedule numbers. Table 5.3 below shows some pipes' sizes most commonly used.

Table 5.3 - Standard Dimensions of Steel Pipe

Nominal In inches	Pipe Outside Diameter in inches(mm)	Pipe Inside Diameter		
		Schedule 40	Schedule 80	Schedule 160
1/8	0.405(10.3)	0.269	0.215	
1/4	0.540(13.7)	0.364	0.302	
3/8	0.675(17.1)	0.493	0.423	
1/2	0.840(21.3)	0.622	0.546	0.466
3/4	1.050(26.6)	0.824	0.742	0.614
1	1.315(33.4)	1.049	0.957	0.815
1 1/4	1.660(42.1)	1.380	1.278	1.160
1 1/2	1.900(48.2)	1.610	1.500	1.338
2	2.375(60.3)	2.067	1.939	1.689

Steel Tubing and Flexible Hoses

In hydraulic system, steel tubing and flexible hoses are the most important final pieces of accessories to the entire installation and delivery of a particular system.

Steel Tubes

Steel tubes are used in hydraulic systems when rigid are required. They are easier to assemble and don't require welding in order to achieve leak-proof connections. Seamless steel tubing is the most widely used conductor type for hydraulic systems as it has significant advantages over pipes. The tubing can be bent into any shape thereby reducing the number of fittings in a system. Tubing is easier to handle and can be reused without any sealing problems. For low-volume systems, tubing can handle the pressure and flow requirements with less bulk and weight. However, the only flip side to tubing and their fittings is that they can be expensive.

Ferrule

Figure 5.2

Steel tubes are measured and specified by their outside diameter and wall thickness. The tubing grade and wall specification determine the pressure ratings. Figure 5.2 shows a component through a tube connector, ferrule and fastening nut. The tube is often pre-flared to an angle of 37° to accept a 37° flare connector.

Some of the most common tube sizes used in hydraulic systems have been tabulated below.

Table 5.4 - OD and ID of Hydraulic Tubes

Tube OD inches	Wall Thickness In mm	Tube ID In mm	Tube OD	Wall Thickness In mm	Tube ID In mm
1/8	0.89	1.5	5/8	0.89	14.0
3/16	0.89	3.0	3/4	1.3	16.5
1/4	0.89	4.5	7/8	1.3	19.8
5/16	0.89	6.2	1	1.65	22.9
3/8	0.89	7.8	1 1/4	1.65	28.5
1/2	0.89	10.9	1 1/2	1.65	34.8

The most widely used material for steel tubing is SAE1010 cold drawn steel. This material has a fairly high tensile strength and is quite easy to work with. In order to obtain higher tensile strengths, the tubes are made of AISI 4130 steel.

Some of these fittings are known as compression fittings. They seal by metal-to-metal contact and may be either a flared type or flare-less type. Other fittings may use O-rings for sealing purposes.

The sleeve inside the nut, supports the tube, to dampen vibrations. When the hydraulic component has straight threaded ports, straight thread type O-ring fittings can be used. This type of sealing is ideal for higher pressures, because as the pressure increases the seal gets tighter.

The tubing is supposed ahead of the ferrules, by the fitting body. Two ferrules grasp tightly around the tube without causing any damage to the tube wall. There is virtually no constriction of the inner wall, ensuring minimum flow restriction.

Flexible hoses

Flexible hoses are one of the most important conductors used in hydraulic systems. They are used in applications where lines must flex or bend or in other words when hydraulic components such as actuators are subjected to movement. Hoses are fabricated in layers of elastomer (synthetic rubber) and braided fabric or braided wire, which permit operation at high pressure.

Hose construction has been standardized by the Society of Automotive Engineers under SAE 15-17 also known as the R series (example being 100 R4) by which the cover, construction, application and pressure rating is described.

There are many suppliers supplying hydraulic hoses with fittings, some well known ones are as follows:

1. Aeroquip

2. Swagelok

3. Parker Hannifin

Chapter 6 - Power System and Controls

Principles of Control System
Basic Electricity
Electrical and Mechanical Energy
Unit of Power
Power Supply
AC and DC Power Generators
Solar Photovoltaic (PV) Panels
Battery Cells
Relays
Solenoids
Controls System
Types of Control Devices
Types of Pressure Switches
Pressure Sensor and Transducers
Temperature Switches and Sensors
Thermocouple

Upon completion of this chapter, you should be able to:

1. Know the basic electrical component for hydraulic controls
2. Know about AC and DC power supply which is vital for control systems
3. The various types of power sources to drive hydraulic systems
4. Know about functions of solenoids and relays
5. Know about the various sensing devices used in hydraulic systems

What you want to be, defines what you become.
Albert Einstein

Principles of Control Systems

Why Control Systems are important to hydraulic system?

We need to emphasize on the fundamentals of process and equipment control. Those affected are the thousands of small builders and users and to some extent the large manufacturers, plus the many system integrators. Most selling organisations, both large and small, include many thousands of manufacturers' agents, distributors, sales engineers, process engineers and maintenance personnel. These are the people charged with the responsibility for selling machines and keeping them operating.

The success or failure of installations may depend on the ability of these people to maintain and troubleshoot equipment properly. The personnel involved not only need to know the whole hosts of software and hardware available to them, they also need to know the relevant knowledge information. With this knowledge, they are in a better position to develop and obtain successful outcomes.

There are six specific areas which we can define and explain the principles of control systems.

1. Components

Components are the building blocks of all systems. The core knowledge of the principles and applications of each component in a system sets the groundwork for a complete understanding of all facets of a control system.

2. Control techniques and circuits

Techniques and circuits are the processes of building the system from the components. Like building a house, without a concept and plan, the structure will fail over time. Techniques and circuits are the lead to a successful plan. The outcome of the control system will depend on how well we acquire it and use them to work productively.

3. Troubleshooting.

Troubleshooting machine control circuits involves locating and properly identifying the nature and magnitude of a fault or error. This fault may be in the circuit designs (electrical or hydraulic), physical wiring or piping, or components and equipment used. The time required and the technique or system used to locate and identify the error is important. Of course, the most importance fact is the time and expense involved to put the machine back into normal operating condition.

4. Maintenance

Preventive maintenance would eliminate the need for most troubleshooting. Many machines are built to last and allowed to operate until they literally fall apart or being replaced by an advance piece.

5. Electrical and hydraulic standards

A reasonable set of standards should be followed. If the intended result is the improvement of design and application to reduce downtime and promote safety, using universal standards can be extremely helpful.

6. Keeping current with changing technology

Keeping current with changing technology means implementing a strategy or designing a system to keep up with time and technology change.

It is not uncommon to have PLC, computer base controllers, RFID, IR, WIFI and or the combinations of all of the above, as latest technologies play a very important role of control systems in modern industries. As the emphasis on productivity intensify, more innovative things will come on stream replacing the old.

Automation is second nature to modern machineries as human moves up the value chains and so would the productivity demand on hydraulic systems and machines.

For example, you can easily have a decent size of a standalone solar PV which pack enough power to operate gate control of your house and if couple with an RFID (via simple network) you could control your gate with a touch on your hand phone/ipad even if you are half way round the globe. It is technology at work.

(NB: We shall discuss wireless control system in chapter 9).

Basic Electricity

Electricity is a fascinating science that we use in many different ways. It would be difficult to think of the many ways that we use electricity each day. It is a very important tool and is vital to have good understanding of electricity and its applications.

Several people have made important discoveries about electricity. Benjamin Franklin is given much credit for the discovery of how electricity can be used. Franklin and other scientists in their times thought that electricity was a force that could have "positive" or "negative" charges.

Today, we know that electricity is produced by the effect of tiny particles called electrons. Electrons are too small to be seen with the naked eye. They exist in all materials. The production of electricity has become an important part of our society.

Basic knowledge of electrical and electronic principles is important working with hydraulic systems. This is because hydraulic systems work from a pump that is operated from an electric motor. Control circuits of the hydraulic systems contain electrical and electronic components. Many hydraulic devices are electrically powered.

There are many electrical components and integrated parts that form a piece of design to translate controls for hydraulic systems. Here, we shall just mentioned two key components that played major role in term of the hydraulic control system. These two electrical components are relays control and solenoids.

We will not cover detail of electric circuitries and controls, as they are topics by itself where you could easily pick one to learn. However, we will cover some basic electricity parameters that are prerequisite for electrical power sizing and calculation purposes. It is also the intention of this book to deliver the minimum electrical concept

Electrical and Mechanical Energy

Current and Voltage Definition

The unit of electric current is the ampere (A). The instrument used for measuring electric current is known as an **ammeter.** The unit of electric quantity is the coulomb, namely the quantity of electricity passing a given point in circuit when a current of 1 ampere is maintained for 1 second.

Hence, Q [coulombs] = I [amperes] x t [seconds]

1 ampere hour (AH) = 3 600 coulombs

(NB: Do not confuse this italic *Q for coulombs* with that of **Q for fluid flow rate,** in chapter 1)

The unit of emf is the volt (V). There is another electrical quantity measured in volts and that is potential difference (p.d). Both emf and p.d. are often referred to simply as voltage. The emf is the voltage of the electrical power supply (in this case the battery). The p.d. is the voltage across the components in the circuit. The resistance R_{int} in Figure 6.1 has a p.d. across its ends.

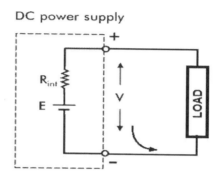

Figure 6.1

The actual work accomplished in a circuit is a result of the potential difference available at the two ends of a conductor. It is this difference of potential that causes electrons to move or flow in a circuit as the one shown in figure 6.1. The potential difference is referred to as electromotive force (emf) or voltage. Voltage is the force that moves the electrons in the circuit.

Resistance

The unit of electric resistance is the ohm (Ω) namely the resistance between two points of a conductor when a potential difference of 1 volt, applied between these points, produces a current of 1 ampere in this conductor. The conductor not being a source of any electromotive force.

Electrical Energy

In deriving the formula for electrical energy or the heat produced by a current, begin with the definition of unit, potential difference (p.d) or volt.

The volt (V) is the difference of electric potential between two points of conductor carrying current of 1 A, when the power dissipated between these two points is equal to 1 W..

When the charge Q is transferred between two points having a p.d. V, the energy W released is given by;

$$W = QV \quad .... \text{(Eq. 6.1)}$$

Where W is in Joules when Q is in coulombs and V in volts.

If the current **I** flow in time t, Q is equal to **I**t,

then, $W = \mathbf{IV}t$

The electric power, P_e = **energy per second**

$$= \frac{energy, W}{time, t}$$

$$\therefore P_e = \mathbf{IV} \quad ....\text{(Eq. 6.2)}$$

It also equal to I^2R (since I = V/R)

where **I** is amperes, **V** in volts, **R** in ohms, then P_e is in Watts (W).

Further reading on basic electricity are recommended and can be acquired fairly easy. (NB: We have another Book on DIY, where you could learn more about electricity)

Mechanical Energy

Mechanical energy is the sum of the an object's potential energy and kinetic energy. Potential energy is how much energy is stored in an object at rest, an example of this form of energy storage in hydraulic application is the accumulator as explained in chapter 5. Kinetic energy is the energy an object has due to its motion.

The SI unit of energy is the Joule (after the English physicist, James P. Joule, 1818-1889). The Joule is the work done when a force of 1 N acts through a distance of 1 m in the direction of the force. Hence, if a force **F** acts through distance d in its own direction,

$$\textbf{Work Done} = \textbf{F [Newtons] x s [metres]}$$

$$= \textbf{Fs Joules} \quad \ldots\ldots(\text{Eq. 6.3})$$

If a body having mass m, in kilograms, is moving with velocity v, in metres/second,

$$\textbf{Kinetic energy} = \textbf{1/2 mv}^2 \textbf{ Joules}\ldots\ldots(\textbf{Eq. 6.4})$$

If a body having mass m, in kilograms, is lifted vertically through height h, in metres, and if g is the gravitational acceleration, in metres/second^2 (m/s^2), in that region,

Potential energy acquired by body = work done in lifting the body

$$\textbf{Potential energy} = \textbf{mgh Joules}$$

$$\textbf{approximately, 9.8 mh} \ldots\ldots(\textbf{Eq. 6.5})$$

Example 6.1

A body having a mass of 30 kg is supported 50m above the ground. What is its potential energy relative to the ground?

If the body is allowed to fall freely, calculate its kinetic energy just before it touches the ground. Assume gravitational acceleration to be 9.8 m/s².

$$\text{Weight of body} = 30 \text{ [kg]} \times 9.8 \text{ m/s}^2$$

$$= 294 \text{ N}$$

$$\therefore \textbf{Potential Energy} = 294 \times 50$$

$$= \textbf{14,700 J}$$

If **v** is the velocity of the body after it has fallen a distance '**s**' with an acceleration **g**, recall one of kinematic formula, $\mathbf{v^2 = u^2 + 2as}$ where u=0 and 'a' the acceleration for free fall situation a = +ve g ;

$$v = \sqrt{2gs}$$

$$= \sqrt{2 \times 9.8 \times 50}$$

$$= 31.3 \text{ m/s}$$

and **kinetic energy** = $\mathbf{1/2 \ m \ v^2}$

$$= 1/2 \times 30 \times 31.3$$

$$= \textbf{14,700 J}$$

Hence, the whole of the potential energy has been converted into kinetic energy. When the body is finally brought to rest by impact with the ground, practically the whole of this kinetic energy is converted into heat.

Unit of Power

Since power is the rate of doing work, it follows that the SI unit of power is the joule/second or watt. In practice, the watt is often found to be inconveniently small and so the kilowatt is frequently used. Similarly, when dealing with a large amount of energy, it is often convenient to express in kilowatt hours rather than in joules.

1 kw.h = 1000 watt hours

=1000 x 3600 watt seconds or joules

= 3 600 000 J or 3.6 MJ

If T is the torque, in newton metres, due to force acting about an axis of rotation, and if η is the speed in revolutions/second,

Power = torque in newton metres x speed in radians/second

So,

$$\text{Power } \mathbf{P_m} = \mathbf{T} \times \mathbf{2\pi} \; \eta \quad \text{joules/second or watts}$$

$$\text{Or } \mathbf{P_m} = \omega \, \mathbf{T} \; ...... \; (\text{Eq. 6.6})$$

where ω *is the angular velocity in* $\frac{radians}{second}$.

Example 6.2

A stone block, having a mass of 120 kg, is hauled 100m in 2 min along a horizontal floor, The coefficient of friction is 0.3. Calculate

(a) the horizontal force required

(b) the work done and

(c) the power

$$\text{Weight of stone} = 120 \text{ kg} \times 9.8$$

$$= 1176 \text{ N}$$

$$\therefore \text{force required} = 0.3 \times 1176$$

$$= 353 \text{ N}$$

$$\text{Work Done} = 353 \times 100$$

$$= 35\,300 \text{ or } 35.3 \text{ KJ}$$

$$\text{Power} = \frac{35\,300}{2 \times 60}$$

$$= \mathbf{294 \text{ W}}$$

Example 6.3

An electric motor is developing 10 kw at a speed of 900 rev/min (rpm). Calculate the torque available at the shaft.

$$\text{Speed} = \frac{900 \, [\frac{rev}{min}]}{60 \, [\frac{s}{min}]}$$

$$= 15 \text{ rev/s}$$

Substituting in expression (Eq. 6.6), we have,

$$10\,000 = T \cdot (2 \times \pi \times 15)$$

Torque T = 106 Nm

In the United States and some countries, mechanical power is also measured in horsepower (hp). This unit is being used from the steam engine age replacing the horses power.

It is the measurement of 1 horse power to lift 550 lbs by 1 ft distant in 1 second. This 1 hp has an equivalent SI unit of about 746 watts.

Power Supply

AC and DC power supply

Alternating current and Direct current

The current from a battery flows from positive to negative terminals, through the circuit that is connected to it. Such a current is called direct current. The battery voltage that produces direct current is called direct voltage. The electrical power that can be obtained from even large batteries is fairly small, hence alternating current (a.c.) is the major driving force in most hydraulic installations.

Alternating Current Supply

For AC current source, also commonly known as AC supply, the current that comes from the main supply source repeatedly changes direction between the 2 terminals. One side of the supply is called live and the other is neutral. The live repeatedly changes from about + 330V to - 330 V. The time taken for the voltage to change from + 330V to - 330 V and back is 1/50 s for 50 Hz and 1/60s for 60 Hz in a cycle also known as the period (T),

where $f = \dfrac{1}{T}$.

Alternating voltage and current are changing the whole time so that it is not easy to give a value to voltage and current. However, there is a kind of 'average' value known as root mean square (rms) which can be generally quoted as,

rms value = 0.71 x peak value (330 V)

which is approximately, 230V

Large size hydraulic pumps are often driven by powerful electric motors know as three-phase induction motors which required 3 phase current. (Refer to appendix II.2 for 3-phase AC system explanation)

Direct Current Supply

Most hydraulic installation uses 12V, 24V and 48V on standalone system, power back-up for safety purposes or as secondary control along with AC supply system. Of these 3 supply voltages, 12V and 24V system are the most common ones.

The most important criterion for battery types selection besides, safety aspect, is the ability to deliver power continuously which is expressed in ampere-hours.

For example, a battery of 12V rated at 100 ampere-hours (i.e. designated by '100'AH) can continuously supply any of the followings:

100 ampere for 1 hour (100A x 1 hr)

10 ampere for 10 hours (10A x 10 hrs)

1 ampere for 100 hours (1A x 100 hrs)

(NB: Usually rechargeable batteries are used in hydraulic installations)

If the DC supply is converted from the main AC supply, then you need not worry about ampere-hours. The only concern is power failure.

AC and DC Power Generators

Magnetism is the most common method of producing electrical energy today. If a wire is passed through a magnetic field, voltage is produced, as long as there is motion between the magnetic field and the conductor. A device based on this principle is called a generator. A generator can produce either direct current (DC) or alternating current (AC) or even both, depending on how it is wired. When electrons flow in only one direction, the current is direct current. When electrons flow in one direction then in the opposite direction, the current is an alternating current. A generator may be powered by steam from nuclear power or coal, water, wind or gasoline or diesel engines.

Commercial Generator sets are many and they could turn fossil fuels into electrical energy efficiently. There are also generator powered by environmental friendly sources like, biofuel, biogas and all other organic wastes.

Most often, you will find them in isolated places where such power generator sets generate the main power supply to drive hydraulic systems. There are examples like on board the ships or vessels, ocean tankers, building and construction machineries, army installations, oil-rigs and drilling platforms.

Solar Photovoltaic (PV) Panels

Sunlight is plentiful and costs nothing, as PV solar panels are cheaper now. It is also free of pollution. For many months in much of the world, sun provides all the thermal energy (heat) people need to stay warm. At times, there is too much solar energy.

If we could harness all the available solar energy, we would have much more energy than we need. The sun provides 1,370 watts per square meter to our outer atmosphere. Some solar energy is reflected into space.

Light energy can be converted directly to electrical energy by light striking a photosensitive (light-sensitive) substance in a photovoltaic cell, also known as a solar cell. A solar cell consists of photosensitive materials mounted between metal contacts. When the surface of the photosensitive material is exposed to light, it dislodges electrons from their orbits around the surface atoms of the material. A single cell can produce a small voltage. Many cells must be linked together to produce a useable voltage and current.

To use it, you need to link the panels in series or parallel to achieve the desire power for your application. For solar energy to provide energy at all times, a solar charger unit and rechargeable batteries to store solar energy must be in place.

Although it is an easy power supply device to install, be very careful about selecting the type of solar PV panel. Most of current commercialized PV devices for consumer electronics just use simple series configurations, such as cell phone chargers, battery maintainers for automobiles or recreational vehicles. The main reason of using series configuration is because it is easy to build up the PV panel output voltage and thus avoiding voltage regulation for low cost. These PV source performance will be degraded particularly under complex illumination conditions.

With significant research and developments of power electronics, power converters/inverters today are highly efficient and low in cost. More of it are integrated into PV generators. It is to be noted for high-voltage applications, many parallel-connected PV modules (rather than cells) that equipped with advanced power electronics have achieved better performance than conventional series-connected PV modules.

Power/Energy storage system - battery

Battery Cells

The second most common method of producing electrical energy today is by the use of a chemical cell. The cell consists of two dissimilar metals, copper and zinc, immersed in a salt, acid or alkaline solution. The metals, copper and zinc are the electrodes. These electrodes are termed as positive and negative separated by an electrolytic solution that provides dc voltage. Batteries and cells are classified as either primary or secondary. Primary cells are those cells that are not rechargeable.

Secondary cells are those that may be discharged and recharged many times. To name a few common ones, the Nickel-cadmium cell, alkaline-manganese dioxide cells, mercuric oxide cells, silver oxide cells, lithium and lithium ion cells

NB: We will not discuss this in detail; however, we will present this topic as a handy guide book on potable batteries unit for energy storage and harnessing solar energy. It is written for those who wish to build their pet projects and for folks who DIY, harnessing and storing green energy.

Relays

A relay, as shown in Figure 6.1, is an electromechanical switch that opens and closes with an electromagnetic coil. The major parts of the relay consist of stationary and moving contacts. The moving contacts are attached to the plunger. Contacts are referred to as normally open (NO) and normally closed (NC). When the coil of the relay is energized, current flows through the coil, it generates a magnetic field that attracts the plunger pulling it down.

This action closes the normally open contacts and opening the normally closed contacts. When the current through the coil stops, a spring pulls the armature back to its original position and opens the switch.

A relay is used where it is desirable to have one circuit control another circuit. It electrically isolates the two circuits. A small voltage or current can control a large voltage or current. A relay can also be used to control several circuits some distance away.

A doorbell is an application of the relay. The striker to ring the bell is attached to the plunger. As the doorbell is pressed, the relay coil is energized, pulling down the plunger and striking the bell. As the plunger moves down, it opens the circuit, de-energizing the relay. The plunger is pulled back by the spring closing the switch contacts, energizing the circuit again, and the cycle repeats until the button is released.

A standard relay consists a coil to generate temporary magnetic attraction. Coils are available to cover most standard voltages from 12V to 600V. Relay coils are typically made of a moulded construction to reduce moisture absorption and increase mechanical strength.

S/N	Descriptions	Physical Appearance
1.	Relay (single pole)	
2.	Relay (multiple poles)	
3.	Relay mounted on PCB	
	NB: Note the lugs on the underside of the relay (1 & 2).	

Fig. 6.1

The two important parts in a relay are the coil and contacts. Of these, the contacts generally require greater consideration in practical circuit and control designs. There is some single-break contacts used in industrial relays. However, most of the relays used in machines control and hydraulic control system have double break contacts. The rating of contacts can sometimes be very misleading. The three ratings generally are:

1. Inrush or "make contact" capacity

2. Normal or continuous carrying capacity

3. The opening or break capacity

For example, a typical industrial relay has the following contact ratings:

1. 10A non-inductive continuous load (AC)

2. 6A inductive load at 120V

3. 60A make and 60A break inductive load at 120V

Generally, most commercial relays will have the following contact information to aid selection purposes for intended functions.

1. SPNO (SPST – NO) – Single pole normally open (Form A)

2. SPCO (SPDT) – Single pole changeover (Form C)

3. DPNO – Double pole normally open (Double Form A)

4. DPCO (DPDT) – Double pole changeover (Double Form C)

Again, it is wise to note that the coils are available in AC or DC types and some even have a light emitting diode (LED) feature or power 'on' indication.

NB: Form A refers to contact switch only with ON or OFF. Form C refers to contact switches with Normally Open (NO) and Normally Close (NC).

While these are usually used in the panel mount inside the enclosure box, also known as control box. There are also some in the form of hermetically sealed and solid state used widely on printed circuit board (PCB). With a sealed construction and no moving parts, these solid state relays are particularly suited to AC switching application requiring long life and high reliability. The switching is silent, causes no arcing and is unaffected by vibration and corrosive atmospheres.

Solenoids

A solenoid consists of a coil in which it house a soft-iron plunger. The plunger normally has just one end of it in the coil. When a current is passed through the coil, the plunger is pulled strongly into the coil.

The general principle of the solenoid action is very important in hydraulic system control. Like the relay and contactor, the solenoid is an electromechanical device. In this device, electrical energy is used to magnetically cause mechanical movement of a plunger. (Refer to figure 6.2 below)
A typical solenoid is made up of three basic parts:

1. Frame
2. Plunger
3. Coil

The frame and plunger are made up of laminations of high-grade silicon steel. The coil is wound with an insulated copper conductor. When the coil of a solenoid is energized, a magnetic field is created about the coil. This magnetic field causes the armature to exert a pulling force on the plunger to move the spool of the valve. The pull forces in solenoids varies widely. It may be as low as 500 gram (1 lb) or as high as 50 kg (100 lbs). Connections to the coil may be supplied in the following ways:

1. Pigtail leads (lugs)
2. Terminals on the coil
3. Terminal blocks
4. Plug-in connections (as in Fig. 6.2)

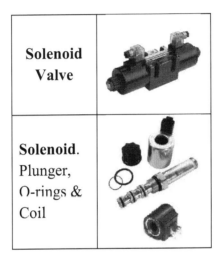

Fig. 6.2

The switching of solenoid valves used in oil hydraulic systems is performed by electrical operations of manual switch, limit switch, electrical relay or pressure switch. Solenoid used includes AC operated type, DC operated type, and AC/DC conversion type.

AC/DC conversion type solenoid functions as AC/DC converter and surge absorber in addition to DC solenoid. It can be directly connected to AC power source. It is characterised by low noise and hydraulic shock during the changeover, and coil does not burn even if the spool stops in the middles of changeover.

Controls System

The words 'controls system' sometimes can be quite misleading by not defining the meaning and specifying the level of controls intended for a particular system. In a simple explanation, it is a set of collective events or activities that could be meaningfully expedited sequentially to perform a particular task. To understand that, as any designer would have to get through first is his or her process thinking. This leads us to the vital decision of choices.

We will explain briefly the main thought process in every aspect of a system to achieve that desired level of controls.

The **first step** is to know the work or function to be performed.

The **second step** is to know the operating conditions under which the starting, stopping and controlling of all or part of the processes to be taken place for the desired outcome or series of outcomes. Practically, all conditions fall into one or more general groups, as affected by;

1. Position

2. Time

3. Pressure

4. Temperature

Each of these conditions can generally be associated with certain intended components. In many cases, the consideration of an actual initiating of a cycle may be through the manually operated push button switch, an external force (not human) initiated the next start process or simply of a time base switches and whatever you could think of.

The **third step** is the most crucial development of the control system. It involves by selecting the most desired control conditions among the various considerations and this completes that 'process thinking task'. There are many times that a control system must be capable of operating under certain sets of conditions to produce the desired results. For example, a control system may be required to operate machine under manual, semiautomatic (single-cycle) or fully automatic (continuously-cycle) operation, so it is the matter of choices, a designer must make.

After a decision is made on which of these types of operations are to be used, then only the components selection can be made. For reasons of safety to both the machine and operating personnel, the machine must operate as desired or intended manner.

Review the three basic steps to start development with elementary diagrams and sketches.

1. Know the work or function to be performed.

2. Know the conditions for starting, controlling, and stopping the process.

3. Finalize the selecting process with the most desired control conditions (outcomes): manual, semiautomatic or automatic.

In designing control circuit or hydraulic circuit for control systems, there are similarities and fundamental problems that should be recognised and overcome it, if progress is to be made.

Some common mistakes in the initial stage of design are:

1. Starting with a circuit that is too large or too complicated.

2. Failing to carry through a mental picture of the component or components into the electrical circuit or hydraulic circuit.

3. Failing to relate physical, mechanical, or environmental actions into devices that convert these actions into useful works.

4. Failing and not performing those actions that will result in damaged components, danger to the operator or machine, or a faulty system and product.

Importance of Pressure Indication and Control

Pressure indication and control are important in many electrical control systems in which air, gas or a liquid is involved. The control takes several forms. It may be used to start or stop a machine on either rising or falling pressure. It may be necessary to know that pressure is being maintained.

The pressure switch is used to transfer information concerning pressure to an electrical circuit. The electrical switch unit can be a single normally open (NO) or normally closed (NC) switch. It can also be a single switch with both NO/NC sharing a common terminal or two independent circuits. Usually, it is easier to use a switch with two independent circuits in the design of a control circuit.

Tolerance in most switches is about 2%; accuracy is about 1%. These are expressed as percentages of the working range. In some cases it is necessary to know and use a differential pressure.

The differential pressure has the range between the actuation point and the re-actuation point. For example, the electrical contacts operate at a preset rising pressure and hold it in place until the pressure drops to a lower level.

This differential may be fixed amount, generally proportional to the operating range, or it can be adjustable.

Type of Control Devices

There are several types of pressure switches available. Some of the most widely used types are:

1. Pressure Switches (Sealed piston, Bourdon tube, Diaphragm and Solid State)

2. Pressure Sensor and Transducers

3. Temperature Switches and Sensors

4. Thermocouple

S/N	Descriptions	Physical Appearance
1.	Pressure Switch	
2.	Pressure Transducer	
3.	Thermocouple (RTD type)	

Fig. 6.2

Sealed Piston Type

The sealed-piston type of pressure switch is actuated by means of a piston assembly that is in direct contact with the hydraulic fluid. The assembly consists of a piston sealed with an O-ring, direct acting on a snap-action switch. An extremely long life can be expected from this switch where mechanical pulsations exceed 25 per minute and where high over-pressures occur. The sealed piston saves the cost of installing return lines.

The piston-type pressure switch can withstand large pressure changes and high overpressures. These units are suitable for pressures in the range of 15 psi to12,000 psi (1 bar to 827 bar).

Type of Pressure Switches

Bourdon-Tube Pressure Switch

The pressure operating element in a Bourdon-tube unit is made up of a tube formed in the shape of an arc. Application of pressure to the tube will cause it to straighten out. Care should be taken not to apply pressures beyond the rating of the unit. Excessive pressures tend to bend the tube beyond its ability to return to its original shape.

The Bourdon-tube pressure switch unit is extremely sensitive. It senses peak pressures; for this reason, it may be necessary to use some form of dampening or snubbing of the pressure entering the tube.

The pressure is available for applications from 50 psi to 18,000 psi (about 3 to 1,241 bar). It is generally used more in indicating service than in switching because of the inherent low work output of the unit.

Diaphragm Pressure Switch

A diaphragm pressure switch consists of a disk with convolutions around the edge. The edge of the disk is fixed within the case of the switch. Pressure is applied against the full area of the diaphragm. The center of the diaphragm opposite the pressure side is free to move and operate the snap-action electrical contacts. Units are available from vacuum of 150 psi (about 10 bar).

Solid State Pressure Switch

A solid-state pressure switch is interchangeable with existing electromechanical pressure switches. Pressure sensing is performed by semiconductor strain gauges with proof pressures up to 15,000 psi (1,034 bar). Switching is accomplished with solid state triacs. An enclosed terminal block allows four different switching configurations:

1. Single-pole, single-throw (NC)
2. Single-pole, single-throw (NO)
3. Single-pole, double-throw
4. Double-make, double-break

Pressure Sensor and Transducers

Pressure Sensor

Pressure sensors are widely used in sensing of pressures for hydraulic and pneumatic applications. The pressure family includes pressure sensors, pressure transmitters, and vacuum sensors. These sensors incorporate the same ceramic pressure-sensing technology found in high-accuracy pressure transmitters. Ceramic pressure sensors are the preferred choice because they can withstand higher overpressure and are more resistant to pressure spikes. The base of a ceramic sensor has a measuring electrode plus a reference electrode.

A diaphragm containing a second measuring electrode is attached to the base with a glass solder layer. The two measuring electrodes form the plates of a capacitor, and with no pressure applied to the sensor, the electrodes are approximately 10 micrometers apart. As pressure is applied to the diaphragm, this distance decreases and thus increases the capacitance value and produces a signal that is proportional to the applied pressure. When combined with its mechanical mounting, this sensing element provides overall long term mechanical stability.

These pressure transmitters have all the features needed for monitoring hydraulic clamping and ram pressures as, for example, those present in injection moulding machines.

Pressure Transducer

With the use of programmable controllers in pressure control applications in industry, the use of pressure has become important.

A pressure transducer utilizes semiconductor strain gauges that are epoxy-bonded to a metal diaphragm. Pressure applied to the diaphragm through the pressure port produces a minute deflection, which introduces strain to the gauges. The strain produces an electric resistance change proportional to the pressure. Usually four strain gauges (or two gauges with fixed resistors) form a Wheatstone bridge.

Temperature Switches and Sensors

Temperature Switches (Thermostat)

Temperature switches using differential expansion of metals may be of two different types. One uses a mechanical link; the other uses a fused bimetal.

The mechanical link has a temperature range of 37.8°C to 816°C (100°F to 1500°F). The bimetal type has temperature range of 4.4°C to 427°C (40°F to 800°F).

The mechanical link type has one metal piece directly connected or subjected to the part where the temperature is to be detected. The metal expands or contracts due to temperature change, producing a mechanical action that operates a switch.

The bimetal type operates on the principle of uneven expansion of two different metals when heated. With two metal bonded together, heat will deform one metal more than the other. The mechanical action resulting from a temperature change is used to operate a switch. In bimetal units the sensing element and switch are generally enclosed in the same container.

The bimetal type is usually the smallest and least expensive type of temperature controller device. It is not suited for high-precision work. However, its compact, rugged construction lends it to many uses. Accuracies of 1% to 5% of full scale can be expected.

Temperature Sensors

Temperature monitoring systems utilizing sensors provide reliable feedback in temperature control applications. Such a system consists of an electronic control monitor and a variety of sensor probes. The control monitor is packaged in a compact stainless-steel housing that can be mounted either directly to a sensor probe or remotely using a cable probe.

The temperature sensor utilizes a PT1000 RTD for temperature measurement (PT100 RTDs can also be used). The PT1000 RTD is a platinum resistor that exhibits a resistance of 1000Ω at 0°C (32°F). Temperature is sensed by measuring the change in resistance of the RTD. Platinum is used because of its good linearity and stability with temperature. As temperature increases, the resistance of the RTD also increases.

The signal from the PT1000 RTD is evaluated by a microprocessor inside the temperature control monitor. This microprocessor along with other electronic components is mounted on a flexible, temperature-stable polyamide film.

Compared to thermocouples, RTDs are more repeatable and more stable. RTDs are also more sensitive, since the voltage drop across the RTD produces a larger signal than a thermocouple. RTDs have better linearity and do not require the cold junction compensation of a thermocouple.

Thermocouple

The thermocouple operates on the principle of joining two dissimilar metals A and B at their extremities. When a temperature difference exists between the two extremity points, a potential is generated in proportion to the temperature difference. It is quite common to find combinations such as copper-constantan, iron-constantan, copper-zinc, and chromel-alumel being used into the thermocouple construction.

Usually, it is used together with a controller unit also known as transducer. The construction of the thermocouple device is shown in Figure 6.2 (3).

Chapter 7 – Basic Hydraulic System

Basic Hydraulic Power System
- Sizing - step by step example
Sizing Hydraulic Cylinder
Sizing Suitable Pump
Oil Reservoir Tank and Pump Size
Electric Motor sizing
Other HPS off the shelves.

Upon completion of this chapter, you should be able to:

1. Perform calculations of actuators needed for projects
2. Design simple hydraulic control system using simple circuit diagram.
3. Define specifications for intended projects.
4. Perform equipment selections procedures.

People forget how fast you did a job, but they remember how well you did it.
Howard Newton

Basic Hydraulic Power System
- Sizing hydraulic system - step by step example

Example - Minimum working pressure

As a general guide, the minimum working pressure for cylinders under no load condition shall not exceed the following values, if the pressure is applied from the head side.

Piston packing shape	Minimum working pressure (shall not exceed either of the following values, whichever is larger)
'O' ring, 'U' packing seal	3 kgf/cm^2 or maximum operating pressure x 6%
'V' packing seal	5 kgf/cm^2 or maximum operating pressure x 9%
Piston ring	1 kgf/cm^2 or maximum operating pressure x 2.5%

1. To calculate the output forces and speed of cylinder

Cylinder output force = Pressure (system) x Cross Sectional Area of cylinder (if piston back pressure is zero).

$$\text{Force} = P \times A$$

$$\mathbf{F_1} = P \times \frac{\pi}{4} D^2$$

$$\mathbf{F_2} = P \times \frac{\pi}{4} (D^2 - d^2)$$

[NB: If there is back pressure,

$$\text{then } \mathbf{F_2} = \{(P - P_{\text{back pressure}}) \times \frac{\pi}{4} (D^2 - d^2)\} \ ]$$

$$\text{Cylinder Speed} = \frac{Amount\ of\ oil\ inflow}{Cross\ sectional\ area\ of\ cylinder}$$

$$V_1 = \frac{Q_1}{A}$$

$$V_1 = \frac{Q_1}{\frac{\pi}{4}D^2}$$

$$V_2 = \frac{Q_2}{\frac{\pi}{4}(D^2 - d^2)}$$

Where,

F_1 : Extension Force of cylinder
F_2 : Retraction Force of cylinder
P : System pressure
D : Inner diameter of cylinder (Bore size)
d : Rod diameter
V_1: Extension speed of rod
V_2 : Retraction speed of rod
Q_1 : Volume flow rate into piston side
Q_2 : Volume flow rate into rod side

Work Example of Hydraulic System

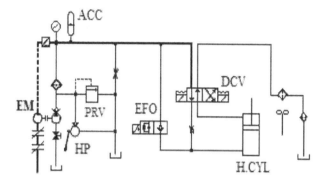

Fig. 7.1 Hydraulic Circuit

Fig 7.1 shows an example of the hydraulic circuit that readers are advised to refer to it as and when require for better understanding of a basic hydraulic system.

Sizing Hydraulic Cylinder

The followings are the usual basic steps to estimate the cylinder load.

In order to simplify the work example, we assume the cylinder task is design to push and pull a load 1 m along it horizontal axis. Please also note that the examples are illustrated in metric unit, and for some, imperials are quoted in brackets for easy references and convenience.

We could compute Load due to frictional resistance if the Load end rest on sliding surface.

We assume static friction $\mu_s \leq 0.2 \sim 0.3$
Dynamic friction $\mu_d \leq 0.05 \sim 0.15$

Although the coefficient of friction varies with condition of contact surface, the coefficient on sliding surfaces of load contact with is generally within the above range.

Force to overcome Frictional load on starting $(F_s) = W \times \mu_s$

Force to maintain Frictional load after starting $(F_d) = W \times \mu_d$

Where, $W = mg$ (g is 9.81 m/s^2)

Points to note:

In practice, to move sliding object, very often rollers are incorporated in the design to take care of frictional load. If roller cannot be introduced, then the contact surfaces are usually smoothen to reduce frictions. This particular exercise, the rollers are omitted to illustrate the frictional influences on the load can be astonishingly high.

When an object say m = 500 kg (approximately 1/2-ton) is accelerated, it is subjected to inertia load which is the product of acceleration and mass of moving object. If the acceleration is constant, the Inertia Force can be expressed by,

$$\textbf{Inertia Force} = m \times a$$

$$\textbf{Acceleration } a = \frac{Changed\ in\ speed}{Time\ required\ for\ acceleration}$$

Assuming also acceleration from 0 to 3 m/min is made in 4 seconds,

$$\text{Then, } a = \frac{3}{60 \times 4}$$

$$= 0.0125 \text{ m/s}^2$$

$$F = m \times a$$

$$\mathbf{F = 6.25 \text{ N}}$$

Force $F_s \sim 1\,470$ N
Force $F_d \sim 740$ N

So the required force of the cylinder required to move m = 500 kg would be,

$F_{total} = (4\,905 + 1\,470 + 740 + 6.2)$

Approximately **7.12 KN**

To size the hydraulic cylinder, it is common practice to look at the force on the retraction stroke, because the area at the rod side (also known as annulus area) is smaller. We assume the intended working pressure is about 150 kgf/cm² (about 2,175 psi), although most cylinders can work to maximum pressure of 200 kgf/cm² (2,900 psi).

To size the cylinder, we have to recall Equation **Eq. 1.1** and example 1.1,

$$\mathbf{F_2 = P \times \frac{\pi}{4}(D^2 - d^2)}$$

NB: It is wise to compute Force F as Kgf (or lb for imperial unit), as most cylinders are standardized with 'cm' for metric (or ins for imperial) measurement.

So, $F_2 = \dfrac{7\,120}{9.81}$

Therefore, F_2 = 730 kgf (after rounding up),

Area of Annulus is,

$$\frac{\pi}{4}(D^2 - d^2) = 12.5 \text{ cm}^2$$

Using the standard cylinder chart (see Table I.2 in Appendix I), showing the pulling force on annulus area working on 150 bar is 1.87 ton. This is about 2.5 times the calculated load needed to slide the object. So it is safe to select Bore size 50mm and Rod size diameter as 30mm. The push force for the same pressure is about 2.94 ton. However, do note that the chart computations are theoretical forces only.

Points to Note:

In practice, it is wise to include other safety factors, works operational factors and duties of the project which designers must bear in mind. In most practical situations, you might end up using the next higher bore size of 63mm with rod diameter of 30mm or next higher one to contain bending load if it exists.

Sizing Suitable Pump

With this information and using the earlier example of 3 m/min (or 300 cm/min), so the volume flow rate that the pump must deliver to the cylinder is as follows;

1. For retraction speed,

    Approximately, 12.5 cm^2 x 300 cm/min or 3, 750 cm/min (3.75 litre/min)

2. For extension speed, if we assume same speed as retraction,

Approximately, 19.6 cm^2 x 300 cm/min or 5, 880 cm^3/min giving 5.88 litre/min (about 358 cu in/min).

This is about 3.92 cc/rev or 0.239 cu in/rev, our estimated pump's geometric displacement and we use this data to **size a suitable pump** for our project.

Next, we can compute the geometric displacement of the pump by using the following formula (refer also to Appendix I for other important formula that you may need).

For metric unit calculation, use

$$\text{Flow rate of pump L/min} = \frac{\textit{Geometric displacement} \left(\frac{cm^3}{rev}\right) X \textit{ Shaft Speed } (rpm)}{1000}$$

For imperial unit calculation, use

$$\text{Flow rate of pump Gal/min} = \frac{\text{Geometric displacement}\left(\frac{in^3}{rev}\right) \times \text{Shaft Speed (rpm)}}{231}$$

NB: Most electrical motors speed are rated at 1,500 rpm, note also the **frequencies use** in your country, it may vary.

From Table 2.2, we will know what are the pumps available for our example. Given that our flow rate is small and not require high pressure delivery in the system, we could then select either a vane or gear pump.

These information are available in the **manufacturers' hydraulic pumps specifications and performance curves**. Examine it closely and select the most appropriate options before finalizing on one set that is suitable for your project. Although this is a small pump sizing exercise, it is the same sizing procedures for larger flow rates and pressure delivery systems.

NB: For pump selection, occasionally it is wise to select the next higher one to that of our calculated one. Some practical issues that you may need to consider or address are flow rate losses due to the number of bends, pipe size including pipe length which is the distance between HPU to actuators, valves selection, flow control valves throttling devices and internal leakages or losses. If these are not your concerns, then it is alright to use the calculated one. Nevertheless, on hindsight, it is wise to check specifications and availability of the pump with the suppliers.

Oil Reservoir tank and Pump size

The oil reservoir forms the largest component of the hydraulic power unit. It is usually a rectangular tank with skid base and forms the structure to which other components are attached. Locate one corner of the tank top cover for a filter breather cap by which oil can be added to the tank. The capacity of the reservoir is based on frequency of use and pump flow rate as explained in chapter 5. For this example we assume a nominal size of 75 liters capacity tank (about 20 US gallons) and select pump size of 4.2 cc/rev (about 0.258 cu in/rev) for intermittent usage and proximity of HPU (Hydraulic Power Unit) and actuator.

The following steps illustrate how we estimate our tank size.

From the above calculation, we know that we need the pump to deliver about 3.92 cc/rev. Using manufacturer pumps' data sheet, we select pump size of 4.2 cc/rev (0.258 cu in/rev) that will deliver about 6.3 l/min (1.67 gpm) or 6.5 l/min after rounding up. Now, if we assume our actuator length of 1 m stroke, we could then estimate the amount of hydraulic oil required and hence, the tank size. (Refer to page 117 on tank size estimation).

1. For Pump, we have 6 times of 6.5L giving 39 L, and

2. For Actuator use (19.6 cm^2 x 100 cm = 1.96L), 3 times of 1.96L giving 5.88L

Total about, 45 L of oil space and depending on distance between HPU and actuator (s), an air space of between 25% to 40% is a practical design that factor into all hydraulic system to buffer the excess of oil flow back to reservoir or for maintenance purposes.

Therefore,

The Total tank size as required for our example including 30 to 40% air space would be,

$$\frac{45}{0.6} = 75 \text{ Litres (about 20 US gallon)}$$

(NB: 1 US gallon approximately 3.785 Litres)

This is also the approximately charge of oil that will be required to fill the lines and hydraulic cylinder. Note also that we only assume one piece of cylinder in this installation. If there are 10 or 20 sets of cylinder of this size, then the oil require is sizeable. Therefore, tank's size will have to increase proportionally.

The tank's level is indicated by a sight glass on its' front face. The reservoir should only be filled with the hydraulic system pressure at zero, otherwise overflowing can occur as a result of oil being displaced out of the accumulator if there is one in the system. The proper oil level is within 1 inch (25mm) of the sight glass top at zero system pressure.

The reservoir tank holds the suction strainer on the pump suction line and also provides the mounting for return line filter. An oil level switch is provided to shut the pump/motor off should oil loss threaten pump failure. A reservoir heater may be required if the ambient temperature so dictates.

Drains are furnished at tank bottom (both sides) for removing water and /or changing fluids and removable cover provided for clean out and access to the components inside. These must be incorporated for maintenance purposes at regular intervals.

In some tanks design, the tank internal is divided with a baffle plate separating the return pipe and suction pipe. This simple device is provided to check sediment, foreign matters and as well pump or pressure line cavitation caused by air bubbles from the return line as discussed in chapter 5.

Gear Pump/Check valve

The gear pump is mounted on a motor adapter and attached to the motor drive shaft by a flexible coupling. The set screws in the coupling halves should be checked for tightness on the pump and motor shafts prior to start up. The pump seals, as are all other HPU component seals, are Buna N. A check valve is located at the pump. Its purpose is to prevent the pressurized oil in the high pressure side of the unit from running back through the pump after the motor shuts off. If it were to fail you would likely notice the motor run backwards and the system pressure fall until zero (pressure gauge indicator).

Another thing to note is, do not start the pump/motor until oil has been put into the reservoir. The pump can only be run dry for a few seconds before damage to the gears and the housing occurs. It is also a good practice to install a shut off valve on the suction line to the pump to facilitate maintenance. This valve must be fully open at all times except when replacing the pump or maintenance. A closed pump shutoff can destroy the pump in seconds.

Motor (EM)

The motor is mounted horizontally and bolted to the HPU framework as well as to the other side of the pump/motor adapter. It is a wise to select a totally enclosed fan cooled (TEFC) design, single or three phase type. The motor voltage and rating is shown on its nameplate; if for 3 phase, note the wiring (refer to Appendix II.2). If in doubt, consult a wiring technician or seek help.

For some other installations, the motor is mounted vertically with the pump set and bell-housing attachment. The pump usually immerged into the oil and is not visible on the HPU top cover. However, you will find control valves and cartridges in sandwich form, just beside the motor set.

Motor/pump direction of rotation is critical. A direction arrow 'mark' is usually marked on the pump. The motor must run in the indicated direction when wiring up. In case if the motor does not run in the proper direction on start up, reversing one or two wires to the control circuit would set it in the right direction.

Electric Motor Power Sizing

Before proceeding to the next stage of calculation, it is wise to do a re-check on the various important parameters for fluid power unit calculation, which apparently is also the electrical power require to drive the hydraulic pump.

After the re-checking procedures are performed and assuming there are no changes to the design, the next crucial step is to consolidate and set design parameters so that we could proceed to size the electric motor power (the work horse for this example).

1. From the above example, with bore size of 50mm cylinder, the area of the piston is 19.6 cm^2 (refer to table I.2 in appendix I).

2. We know the linear speed of the cylinder rod to be moving at

3 m/min (retracting speed), with this we need about 6 l/minute of hydraulic fluid.

3. The system pressure that the pump must supply is 150 bar. Again in hydraulic, there are certain unintentional power losses somewhere in the system. So we shall use 200 bar in our calculation, incidentally that is also most small pumps especially gear and vane pumps could attain.

(NB: For piston pumps and other higher ends, are about 350 or 400 bar, it is wise to refer to the manufacturer's pump performance curve and specifications sheet).

Using formula,

$$\text{Hydraulic power} = \frac{Flow\ Rate\ (lpm)\ x\ Pressure\ (bar)}{600}$$

(The unit is in Kw for metric unit) Eqn.7.1
or,

$$\text{Hydraulic power} = \frac{Flow\ rate\ (gpm) x\ Pressure\ (psi)}{1714}$$

(The unit is in Hp for imperial unit) ....Eqn. 7.1.1

From the above formula, if we are using 150 bar, we need 1.5 Kw driving power.

If we need system pressure to run at 200 bar (2,900 psi), the motor drive must be at least 2 Kw (or 2.68 hp) theoretically. Please note that there is no standard electric motor with 2.68 hp, so it is always wise to select the next higher one. In this case, our example requires a 3 hp electric motor.

Alternatively, we could also use a lower power electrical motor with an accumulator. With the right components combination and calculations, it is always possible to save on electrical motor size, especially for larger size motor power and of course, if the cost justify using the accumulator combination.

Accumulator (ACC)

The accumulator is a cylindrical pressure vessel that provides the high pressure reserve of oil used to move the cylinders and keep them in position (Refer to chapter 5 for the various type of accumulators). In addition, the oil stored in the accumulator is available to move the cylinders even if the pump/motor should be inoperable. The amount of oil directed out to the cylinders is not limited by the displacement rate of the hydraulic gear pump but by the oil stored in the accumulator.

An accumulator is always size to cater for at least 2 to 3 times of oil demand for activating and deactivating the cylinders should there be an interrupted power failure. Of course, many more times are always better but you have to justify economic sense and not to oversize excessively.

Weight, sizes and safety aspects are important parameters to consider if you need to work with accumulators.

Pressure Switch

The pressure at which the oil side is maintained is determined by a pressure switch mounted on the high pressure (between pump or accumulator) side of the system. As an example we shall select one that is set for proper shutoff pressure of 131 bar (1900 psi) and has 34 bar (500 psi) 'dead-band'. This means that the pressure will fall approximately 34 bar (500 psi) after shutoff about 97 bar (1400 psi) before the switch closes to restart the pump motor. These settings is only an example and provide a general guide. You should also note that too narrow a 'dead-band' would result in undesirable start-stop to the electrical motor. Nevertheless, you could easily select one that suits your application and system.

Pressure Gauge

A pressure gauge is provided to indicate the hydraulic oil pressure of the system. It does not indicate the accumulator pre-charge pressure. The gauge is liquid filled with glycol or glycerine to eliminate needle bounce and a vent is thus provided to allow the case to breath, preventing case blow out.

This gauge must be installed in the HPU pressure lines and should read zero before any pressure introduction. Otherwise, it is not registering or measuring the correct pressure of the hydraulic system. Sometimes, a gentle tapping on the gauge glass will provide the most accurate readings.

Pressure Relief Valve (PRV)

A pressure relief valve is provided, a compulsory valve should the high pressure switch fail to shut off the pump motor. The relief valve is typically set 13.8 to 17 bar (200 to 250 psi) **higher than** the high pressure switch. When the pressure relief valve opens, oil is allowed to circulate from the pressure side of the system to the tank/motor suction. The motor horse power is thus being turned into heat across this valve which could cause component damage if allowed to operate uncorrected.

An open pressure relief valve will cause a hissing noise and if the motor is not running, a falling pressure gauge indication would be noted.

Points to note

The pressure relief valve should in no case be set higher than 1.1 times the pressure rating of the minimum rated component of the whole system. Please note that most components are designed with a 4 to 1 safety factor, thus the burst pressure of a 172 barg (2,500 psig) rated hose would be 689 barg (10,000 psig).

Low level Switch

As noted above, an oil reservoir low level switch is provided to shut down the pump/motor if the reservoir level drops to the point where the suction of the pump could become uncovered.

At this point, we have to emphasize that the gear pump (or most hydraulic pumps) can only run dry for a few seconds before severe wear occurs on the gears and its housing. Causes of low level are slow system leaks and catastrophic failure of the pressure lines or hoses.

Oil Filter

A return filter element may be furnished to filter the oil as it is being returned to the oil reservoir. The oil filter housing is only rated at10 bar (150 psig) or less as the oil in the return line has only to overcome the pressure drop through the filter itself. If the filter should become clogged with dirt from the system a bypass check valve inside the filter will open and allow the dirty oil to circulate back to the reservoir. For this reason regular filter maintenance is a must 'do'.

Directional Control Valve (DCV)

A solenoid actuated directional control valve is provided to direct the high pressure oil to the extension or retraction side of the cylinders. One or more (depending on the number of cylinders to be controlled) are mounted on an aluminium manifold bolted to the back plate on the oil reservoir. When the 'extension' side is energized, the valve connects the high pressure P side of the manifold to the (B) outport port of the manifold. The tank return line (T) is simultaneously connected to the (A) output port. When the 'retraction' side is energized, the manifold (P) side is connected to the (A) port and the (T) side is connected to the (B) port.

The directional valve is equipped with pin extensions mounted on the solenoid ends (this is for the manual override – described above) so that the valve spool can be manually shifted by inserting a pin with a diameter of approximately 3mm (0.125 inch). An Allan key of the same size can easily perform this operation.

Emergency Fast Operate (EFO) valve (optional)

Sometimes a system may be equipped with an optional emergency fast operate (EFO) bypass valves. In this exercise, this valve is shown in Figure 7.1 above, it is located just to the left side of the directional control valve. It is also a solenoid driven valve. When energized directly it immediately connect the high pressure (P) side of the HPU to the 'extension' side of the cylinder(s). This bypasses the normal speed control valves and allow the cylinder to extend at the maximum possible speed. This valve deliberately allows dual speeds operation and control.

Speed Control Valves (Flow control)

Each directional valve station has speed control valves to control the normal extension and retraction speed of the cylinders. It is located in the B line. It is usual practice to use one of this flow control valve to control both the extension and retraction speeds for some practical situations (as described in chapter 3).

Clockwise turning of the adjustment knob is slower (valve closing), faster speed is gained by opening the valve (counter-clockwise). The valve should be locked with the set screw provided after adjustment. This locking is necessary as vibration will dislodge adjusted position giving unintended result.

Hand Pump (HP)

A hand pump is a mechanical device sometimes too important to be omitted in any HPU design. It become very handy should power be lost to the pump/motor or accumulator due to power failure. It can be used to charge up the accumulator pressure fairly quickly or raise/lower cylinders by simply working on a manual hand pump. It is usually mounted adjacent to the pump/motor on the skid base. The hand pump has its' own internal check valve so no fluid is lost through the hand pump back to tank during normal motor drive pump operation.

Ideally the hand pump should be located near the reservoir bottom to run a short suction line. In use, the hand pump supplies oil to the pressure (P) side of the hydraulic system.

To raise the cylinders with the hand pump when electricity is out:

1. Check sight gauge for proper fluid level, add oil as necessary.

2. Make sure the accumulator bypass (bleed down) valve is closed.
3. Shift directional valve spool of cylinders to the 'UP' side or 'Down' side as appropriate.
4. Start pumping (each stroke should be productive), otherwise re-check 1 to 3 again.

System Bleed Down Valve

System bleed down valve is a necessary piece of valve that must be included as part of tank accessories. It can be just a single piece for a simple installation or handful of it for more complex control systems. It is usually located above the tank reservoir. It major functions are described below.

1. It provide a pressure stop valve and sometimes it is especially necessary to bleed the power unit down to zero hydraulic power before replenishing the reservoir with fresh oil; large oil spillage can occur if the unit is not at zero pressure when the reservoir is topped up!

2. It is also installed with accumulator as bypass or bleeds down needle valve located between the high pressure side of the system and the reservoir tank.

3. It is a common practice to mount this type of valve behind the hand pump in a line tied (meaning a 'T' joint) to the hand pump suction line, thus isolating it from the main system when not in use.

So by putting the various components together, our hydraulic system in hydraulic circuit diagram would look somewhat like in figure 7.1. Notice that there are some components like, oil cooler (for hot regions or heavy usage), heater (for cold regions) and fans in the diagram are for specific needs which are omitted in this exercise.

Other HPUs off the shelves.

There are manufacturers and suppliers who supply ready make HPUs. These HPUs of various sizes usually consist of the basic components including motors. We shall just highlight some of these examples as in figure 7.2 for reference.

S/N	Ready Make HPU	Descriptions
1.		This is a low duty hydraulic power unit with very limited function. The power is usually fractional HP. Some may be made as high as 1 HP. The main advantages are small, compact and portable.
2.		This power unit is also a low end power unit. As you can see there is a solenoid DCV on a manifold block. Note the tank capacity when selecting this ready make HPU.
3.		This power unit is quite common in many mid range installations, whereby there are adequate top space, you could add on stack control valves to control several actuators. Again, you have to note about tank size and also the HP size of the electric motor.

Figure 7.2

Chapter 8 – Design Procedure and Troubleshooting

Hydraulic System & Equipment Design Procedures
Hydraulic System Troubleshooting

Upon completion of this chapter, you should be able to:
1. Acquire the hydraulic components for a particular hydraulic system confidently.
2. Setup a hydraulic system and it control system confidently.
3. Perform basic troubleshooting if you encounter hydraulic problems.
4. Perform components inspection
5. Perform site commissioning of hydraulic system

If you wait until you have to change, you may have waited too long.
Jack Welch

Example of Hydraulic equipment design procedure

Steps	Descriptions
1.	Establishing purposes of operating the machine
2.	Setting the cycle time
3.	Establishing environment/working conditions, accuracy and economy
4.	Determining the circuit pressure (or system working pressure)
5.	Designing the hydraulic circuit
6.	Selecting the actuator – hydraulic cylinder or motor
7.	Selecting the hydraulic pump
8.	Selecting the hydraulic control valve
9.	Determining the pipe size
10.	Determining the tank capacity
11.	Designing the hydraulic equipment

Hydraulic System Troubleshooting (Basic)

This troubleshooting guide is designed to assist service technicians when diagnosing hydraulic system for operational discrepancies. It's important for the engineer or technician to apply these procedures in step by step approach in order to determine the cause of a malfunctioning hydraulic circuit. This guide will reduce the probability of error by the process of elimination.

Troubleshooting instruments

Hydraulic systems depend on proper flow and pressure from the pump to provide the necessary actuator motion for producing useful work. Hence, the measurement of flow and pressure are two important parameters of troubleshooting a hydraulic system Temperature is the third important parameter measured periodically as it affects the viscosity of oil. The use of flow meters can help in determining whether or not the pump is producing the proper flow.

No oil discharge from the pump

Possible causes;

1. The motor rotates in the wrong direction
Action: stop the motor immediately; reverse the electrical wiring to the motor.

2. Oil level in the tank is too low
Action: add oil until oil-level is visible on sight gauge.

3. Air enters in the suction pipe system
Action: locate the leak and repair.

4. Viscosity of oil is too high for the particular pump
Action: check pump's specifications.

5. Pump shaft rotation is too slow for oil suction
Action: check minimum speed recommended by pump's specifications.

No pressure produced

Possible Causes;

1. Set pressure for the relief valve is too low.
Action: Adjust the relief valve by turning, usually a knob to increase the spring tension hence increase the pressure setting.

2. Oil flowing in the pressure supply system returns to the tank freely.
Action: Check each system.

3. Oil flowing in the hydraulic pipe control system returns to the tank freely.
Action: check if each valve operates normally, and if not, make the adjustment. Check also the spool type of directional control valve is correct type.

Noise generated from the pump.

1. Air enters through the joint between the pump and suction pipe.

Action: Apply sealant or other form of gasket (e.g. o-rings, Teflon tapes wherever appropriate to the type of fittings) to the joint. This should eliminate the problem.

2. Air bubbles in oil
Action: check if the return pipe is in the oil and far from the suction pipe. Some tanks have 2 sections specifically divided for suction and return pipes.

3. Viscosity of oil is too high
Action: Use recommended oil for the particular pump

4. The filtration device at suction side is too small
Action: Replace with a large one.

5. The suction pipe of filtration device is clogged.
Action: Clean the filtration device immediately.

Unusual heat produced

1. Viscosity of oil is too high.
Acton: check if the oil is old or it has adequate viscosity

2. Internal leak is excessively large.
Action: check for wear or loosening of packing

3. Discharge pressure is too large.
Action: check the pressure gauge, and set the relief valve at a level required for maximum load.

4. The oil cooler does not work

Action: check if oil is bypassing the oil cooler system or if cooling water flow rate is operating correctly.

5. Air lock in the hydraulic system while installation or re-installation of piping or hoses.

6. The phenomena of cavitations and slippage in pumps will also generate excessive amount of heat.

Action: check air lock at hoses and hydraulic cylinders by purging procedures. After the purging, cylinder will stroke in and out smoothly.

7. A misaligned coupling that results in excessive load on the bearings generating heat. Misalignment of coupling can wear both pump and motor driving shafts prematurely.

Action: check or change coupling immediately.

8. An unusually warmer return line that could be due to operation at relief valve setting.

Action: Check relief valve setting.

9. Increase in the internal leakage of components due to the use of low-viscosity hydraulic fluids and leading to heat generation.

Action: Change hydraulic fluid to the correct viscosity type of fluid.

Points to Note:

Purging procedures

To purge air lock trap in the hoses, loosen the swivel joint slightly at the cylinder port (extend port end). Activate valve to extend cylinder by inching, you will notice air will squeeze out with some oil flowing out.

At this point, tighten the swivel joint and the cylinder continue outward until it entire stroke. Repeat the procedure by loosening the swivel joint slightly at the front port (return port end). You perform this action until the cylinder stroking in and out **without jerking**, then it is well done.

Commissioning Procedures

Incorrect commissioning of hydraulic components during initial start-up can result in damage due to inadequate lubrication, cavitations and air-lock that may not manifest itself as intended for it services. To avoid damage to the system components during start-up, the manufacturer's commissioning procedures should be followed wherever available.

The following are general procedures for commissioning hydraulic systems after components have been replaced from the system or if any other maintenance action has been carried out. The same procedures can be applied when commissioning new system.

Pre-Commissioning

If the system is down as a result of a major component failure, such as a pump failure:

1. Drain and clean the reservoir to ensure that it is free from metallic debris and other contamination. (NB: Take note also the precaution mentioned in chapter 3 prior to executing this step)

2. Change filters.

3. Change the fluid. In large systems where the cost of changing the fluid may be prohibitive, the fluid should be circulated through a 10 μ filter (without bypass) before recharging it into the system.

4. When fitting pumps and motors, check the drive coupling for its alignment with the pump shaft.

5. On closed loop systems, closely inspect the high-pressure hoses or pipes and replace any suspect lines. A blown hose or a pipe can destroy a pump or a motor through cavitation.

6. After fitting each cylinder, fill the cylinder with clean oil wherever possible, through its service port before tightening the swivel joint of the service lines or hoses. (chapter 3 described steps briefly which may be needed). This is to ensure that air remain trap inside the system, which may result in damage to the seals or the cylinder itself.

7. After fitting motor and other connecting lines: In case of piston type motors, fill the motor casing with clean oil from the upper most port and connect the casing drain line.

8. After fitting the pumps and connecting the service lines: open the suction line valve at the reservoir and vent out all the air from the system at the pump suction line.

Commissioning

1. Check that all the pipe and hose connections are tight.

2. Confirm that the reservoir fluid level is above the minimum level.

3. Confirm that all the controls are in neutral so that the system will start in an unloaded condition. Take safety precautions to prevent machine movement when the system is activated during initial start-up.

4. Where the prime mover is an electrical equipment, confirm whether the direction of rotation of the motor is correct by inching the motor.

5. Start the prime mover and run at the lowest possible speed (rpm).

6. On variable displacement pumps and motors with external, low pressure pilot lines, vent out the air from the pilot line and ensure that the line is full of oil. (NB: Do not bleed the pilot lines carrying high-pressure fluid - personal injury may result. If in doubt do not bleed pilot lines.)

7. Allow the system to run in idle mode and unloaded for 5 min. Monitor the pump for any unusual noises or vibrations and inspect the system for leaks and observe the reservoir fluid level.

8. Operate the system without a load. Stroke the cylinders slowly, taking care not to develop pressure at the end of the stroke to avoid compression of trapped air, which may result in damage to the seals. Continue to operate in this manner until all the air is expelled and the actuators operate smoothly.

9. With the system at the correct operating temperature, check and if necessary adjust pressure settings according to the manufacturer's specifications.

10. Test the operation of the system with appropriate load.

11. Inspect the system for leaks.

12. Shut down the prime mover. Remove all test gauges fitted for purposes of monitoring during commissioning (leave the one fitted for system pressure line) and check the reservoir fluid level and top-up if necessary.

13. Put the machine back into service.

Chapter 9 - Advance Technology update

Data and Digital Age
Industrial Networks
Ethernet and the Information Highway
Data Transfer Rate
Open System and Proprietary System
Wireless and Mobility

Upon completion of this chapter, you should be able to:

1. Know about data and control.
2. Know the important of local area network (LAN)
3. Know about bit and byte where data streaming, that is how computer and machine communicate each other to perform certain tasks.
4. Know about wireless, internet and understand how we can use it to our advantage and become masters of technology.

Once a new technology rolls over you, if you're not part of the steam-roller, you're part of the road.
Steward Brand

Introduction

The important about technology update

The world is changing rapidly, and technology is the engine for that change. A nation lacking 'the leading technology' will have a lot to catch up and depend on others that are in the lead. Rapid change must also be understood, or the technology may control us rather than we are controlling the technology. In order to contribute most efficiently to our technology we must be ready for the challenge of learning a seemingly insurmountable amount of information, often based on a good understanding of the math-sciences.

Like the quip in the heading said whether you are part of the roller or the road, You must be aware of the ways technology has already improved our environment.

Change creates opportunities for those who prepare themselves with the skills, knowledge, and attitudes to solve problems. This chapter is about technology, skills, knowledge, and attitudes possessed by the technologists who live and work in a world where "the only encounter is (or constant) change".

The internet as we already know have great impact into our lives how it enables us to acquire information and use it to our advantage. This is part of technology awareness that we must take notice of from now on, if we shall ask ourselves "how could we continue to stay competitive and still be relevant over time, space?"

Data and Digital

Data and digital are 'languages' recognized by the computer. Industries relies on production processes and machines that produce a great deal of data. These data are used to create information such as product quality and production results. Some of the factors that necessitate the conversion of industries into so-called distributed data factories include:

1. Production systems are becoming integrated; in other words,

complete manufacturing operation may be composed of several individual processes.

2. Processes are dependent on other processes for information to complete their objectives.

3. Processes are controlled by any number of individual controllers (PLCs).

4. Sensors and actuators are becoming smarter. They produce computer digital signals that can communicate directly with a controller.

5. The amount of data created in the factory, and electronically processed into control and management information. As such, it has broadened the term control technology to information technology (IT).

The distributed data factory becomes a complex set of computer-controlled processes. Like building architects, engineers develop the unique design of the interrelated control systems to create an architecture perspective to information technology. This architecture is a road map illustrating how the individual sensors, actuators, displays, and controllers are combined to form a distributed data factory.

We are familiar with the control functions of a PLC but this chapter covers the communication tasks where the PLC can perform with sensors, actuators, other PLCs, and business computers. Effective communication between machines and other equipment is essential for coordinating the functions of automation system such as computer-aided manufacturing (CAM), computer-integrated manufacturing (CIM) and control area networking (CAN).

Industrial Networks

Networks are created to move a variety of data from many different sources to many different destinations. For distributed data applications in industry, data from one source may be needed by several controllers. The output of a controller may be sent to an actuator and as input data to another controller. The only practical solution is to have a network of switched data lines that can be controlled at the appropriate time and place to meet the needs of the process.

A network is defined as the interconnection of computers, controllers, and I/O devices to form a path over which to communicate and share data and information. The network circuit is made up of cables, switches and routers. Each device on the network, called a node, has its own unique network address and has the ability to send and receive data on the network as directed by the controller and network software. The assigned unique address is the key to each device being able to recognized as the source or destination of the data. For many factory automation open-system architectures, the address of the device is automatically assigned as units are connected to the network.

The ability for the data to know where they are to go once they are on the network is determined by the switching method used by the network circuit. The network circuit devices recognize the source and destination addresses and makes the correct circuit routes to ensure that the data are moved to the right location.

Ethernet and the Information Highway

The Ethernet protocol is considered to be a stable and reliable protocol for common local area networks (LANs). Many business offices have adopted Ethernet as their LAN protocol. Therefore, when industrial controllers are sending information to a computer used in the business office, the network segment from the controller to the business computer uses Ethernet. Ethernet uses a bus or star topology.

For companies having several factories and the need to control the data between all factories, the network becomes a wide area network (WAN). Currently, the Transmission Control protocol or Internet protocol (TCP/IP) is widely used. This protocol is the standard for the internet. Data utilizing this protocol can be transported to a server connected to the internet.

The following are some examples of PLCs/via LAN or wireless I/O cards and modules for specific applications. They are mostly open sources, it is almost like plug and play, readily available in the market.

With the abundant sources of modules readily available in the internet sites, you could simply find one that is suitable for your project. This has deliberately speed up design and development process, which is unthinkable just decade ago. Of course, now your job is to ensure it has the least compatibility problems and that they have ongoing support and development.

For folk who has no programming skills, you could easily sub-contract the programming part.

NB: Most modules work on Windows, Mac OSX and Linux system.

S/N	Examples of PLC I/O controller modules
1.	
2.	
3.	

Fig. 9.1

DATA Transfer Rate

The unit of measure for network data transfer rate is bits per second (bps or b/s). Even though the rated transfer rate of a network is very high, the typical bit rate is much lower. For example, at the device level the transfer rate may be less than 1 million bps (Mbps). At the control level the transfer rate may be 5 million bps (5 megabits per second or 5Mpps). At this information level, the transfer rate may be rated at 10 Mbps to 100 Mbps.

Open Systems and Proprietary System

The electronics, which control data networks, are a mix of computer systems, some of which follow industry standards (called open systems). While others are proprietary (unique) systems designed by a specific company.

Proprietary systems are constructed of data equipment following the same proprietary software and hardware. The benefit of factory data communication network architecture using proprietary network equipment is that all of the computer systems within the factory can be connected to the same network and the transmission of information is controlled through specific software. However, proprietary systems are only available from a specific manufacturer.

The benefit of open-system architecture is that the data communication equipment can be purchased from several different manufacturers. However, incompatibility between equipment is a problem. It is a small problem the technologist must solve or tweak and technicians must maintain. To most open-system users, the benefits are more than offsetting the little inconvenience.

Wireless and Mobility

Few technologies impact our lives like wireless, because few enable the one basic human element so highly regarded above nearly all others: the freedom to move.

It is common for us to purchase items from stores we never see by shopping online using laptop and desktop computer instead of cars.

The internet has changed the very definition of a store from a bricks-and-mortar establishment only to include a warehouse the end customer never sees, and even warehouses that aren't seen by anyone because they are virtual. Wireless mobility evolves the experience one generation further by un-tethering the shopping experience. Instead of using a desktop or even laptop computer from home or the office to make online purchases, we now can use ipad or smart phone from anywhere in the globe where wireless connectivity is available.

Distances are being handled with improving ease over the past decades as the supply chain of commerce has increased enormously in speed and capacity. Technology is now relatively evenly interspersed along the entire supply. The impact of wireless mobility in this continuum? Distance is now irrelevant to all, intellectually speaking we are all "connected".

Within the enterprise-class networking realm, the IEEE 802.11 standard is the undisputed king of broadband unlicensed wireless in terms of total chipsets deployed. Broadband in this context is a device capable of routinely delivering in excess of 1 Megabit per second (Mbps). However, the numerous other unlicensed wireless protocols are being deployed in today's networks, such as ZigBee, Bluetooth, HART, Trilogy, and many others.

In the realm outside of enterprise-class networking, devices based on these protocols have shipped in much greater quantities than devices based on 802.11. It's noteworthy, then, that the two behemoths of wireless mobility, 802.11 and mobile cellular, are beginning to converge. Bluetooth and mobile cellular converged onto common platform several year ago. If a "network device" is defined as anything that carries voice, video, or data, then mobile cellular, Bluetooth, and other wireless protocols have their rightful place in networking.

Networking is no longer the express unlicensed wireless domain for 802.11; and indeed it may never have been. For example, today's networks in many facilities routinely have an enormous number of different wireless protocols and /or spectrums in use, including those in the following list, which should be considered only representative, because there are doubtless other spectrums and protocols transmitted and received at major facilities:

1. 802.11
2. ZigBee
3. 802.15
4. Mobile cellular in 850, 1700, 1900, 2100 MHz
5. Bluetooth
6. Two-way walkie-Talkie
7. The very familiar one like car remote door locking system

Appendices - Important information

Appendix I

I.1 Conversion Factors

Table I.1 - Pressure & Liquid Head

PSI	KPa	Kgf/cm²	Atmos Or Bar	Ft of water	Metre of water	mmHg	inHg
1	6.895	0.0703	0.68	2.31	0.704	51.87	2.042
0.145	1	0.0102	0.01	0.335	0.102	7.52	0.296
14.22	98.09	1	0.98	32.85	10	737.9	29.05
14.5	100	1.02	1	33.50	10.21	752.1	29.61
0.433	2.986	0.0305	0.03	1	0.305	22.45	0.884
1.42	9.797	0.1	0.0.98	3.281	1	73.66	2.90
0.019	0.131	0.0014	0.0013	0.044	0.014	1	0.039
0.491	3.377	0.0345	0.0339	1.134	0.345	25.4	1

NB: 1 mm Hg is also known by the name 'torr'. (1 kpa = 1 kN/m²) The international standard atmosphere (1 atm) = 101 325 pascals or 1.013 25 bar. This is equal to 1.033 23 kgf/cm² or 14.6959 lbf/in². In meteorology 1 millibar = 100 pascals (1 mb = 100 Pa).

Table I.2 - Pressure/Force Table of standard hone tube cylinders

Sizes		Area			Theoretical Forces					
					Pushing Force on Piston Area			Pulling Force on Annulus Area		
Piston Dia	Rod Dia	Piston Area	Rod Area	Annulus Area	150 bar	200 bar	250 bar	150 bar	200 bar	250 bar
mm	mm	cm²	cm²	cm²	ton	ton	ton	ton	ton	ton
40	20	12.6	3.1	9.5	1.89	2.52	3.15	1.43	1.90	2.38
	25		4.9	7.7				1.16	1.54	1.93
50	25	19.6	4.9	14.7	2.94	2.93	4.91	2.21	2.94	3.68
	30		7.1	12.5				1.87	2.50	3.13
63	30	31.2	7.1	24.1	4.68	6.24	7.80	3.62	4.82	6.03
	35		9.6	21.6				3.24	4.32	5.40
80	40	50.2	12.6	37.7	7.53	10.0	12.6	5.66	7.54	9.43
	50		19.6	30.6				4.59	6.12	7.65
100	50	78.5	19.6	58.9	11.8	15.7	19.6	8.84	11.8	14.7
	70		38.5	40.0				6.0	8.0	10.0
125	70	122.7	38.5	84.2	18.4	24.5	30.7	12.6	16.8	21.1
	90		63.6	59.1				8.87	11.8	14.8
140	70	153.9	38.5	115.4	23.1	30.8	38.5	17.3	23.1	28.9
	100		78.5	75.4				11.3	15.1	18.9
150	70	176.7	38.5	138.2	26.5	35.3	44.2	20.7	27.6	34.5
	100		78.5	98.2				14.7	19.6	24.6
160	80	201.1	50.3	150.8	30.2	40.2	50.3	22.6	30.2	37.7
	100		78.5	122.6				18.4	24.5	30.7
180	90	254.5	63.6	190.9	38.2	50.9	63.6	28.6	38.2	47.7
	125		122.7	131.8				19.8	26.4	33.0
200	100	314.2	78.5	235.7	47.1	62.8	78.8	35.4	47.1	58.9
	140		154.0	160.2				24.0	32.0	40.1
220	125	380.1	122.7	257.4	57.0	76.0	95.0	38.6	51.5	64.4
	140		154.0	226.1				33.9	45.2	56.5
250	125	490.9	122.7	367.3	73.6	98.2	123.0	55.1	73.5	91.8
	150		176.7	314.2				47.1	62.8	78.5

I.2 Common Units Conversion Factors

Conversion Factors

1 m	• 3.281 ft
1 mm	• 0.0394 in
1 cm3	• 0.061 in3
1 Litre	• 1,000 cc
1 Litre	• 0.264 US Gallon
1 LPM	• 0.264 gpm (US)
1 kg	• 2.205 lb
1 bar	• 14.5 psi
1 bar	• 1.02 kg/cm2
1 Nm	• 8.85 lb-in
1 Nm	• 0.102 kg.m
1 kw	• 1.341 hp
1 kw	• 3,412 Btu/hr

I.3. Output forces and speed of cylinder formulae

Cylinder output force = Pressure (system) x Cross Sectional Area of cylinder (if piston back pressure is zero).

$$\text{Force} = P \times A$$

$$F_1 = P \times \frac{\pi}{4} D^2$$

$$F_2 = P \times \frac{\pi}{4}(D^2 - d^2)$$

$$\text{Cylinder Speed} = \frac{Amount\ of\ oil\ inflow}{Cross\ sectional\ area\ of\ cylinder}$$

$$V_1 = \frac{Q_1}{A}$$

$$V_1 = \frac{Q_1}{\frac{\pi}{4}D^2}$$

$$V_2 = \frac{Q_2}{\frac{\pi}{4}(D^2 - d^2)}$$

Where,
  F_1 : Extension Force of cylinder
  F_2 : Retraction Force of cylinder
  P : System pressure
  D : Inner diameter of cylinder (Bore size)
  d : Rod diameter
  V_1: Extension speed of rod
  V_2 : Retraction speed of rod
  Q_1 : Volume flow rate into piston side
  Q_2 : Volume flow rate into rod side

I.4 Useful Hydraulic Formulae

Shaft Power

$$KW = \frac{Torque\ at\ shaft\ (Nm) \times shaft\ speed\ (rpm)}{9{,}550}$$

$$HP = \frac{Torque\ on\ Shaft\ (lb-in) \times shaft\ speed\ (rpm)}{63{,}025}$$

Geometric Flow Rate (Cylinders)

$$L/min = \frac{Effective\ Area\ (cm^2) \times Piston\ Speed\ (\frac{m}{min})}{10}$$

$$gpm = \frac{Effective\ Area\ (in^2) \times Piston\ Speed\ (\frac{in}{min})}{231}$$

Theoretical Thrust (Cylinders)

$$N = Effective\ Area\ (cm^2) \times Pressure\ (bar) \times 10$$

$$lb = Effective\ Area\ (in^2) \times Pressure\ (psi)$$

Velocity Of Fluid In Pipe

$$m/sec = \frac{Flow\ Rate\ (\frac{l}{min}) \times 21.22}{D^2}$$

Where D = inside diameter of pipe in mm.

$$\text{ft/sec} = \frac{Flow\ Rate\ (gpm) \times 0.408}{D^2}$$

Where D = inside diameter of pipe in inch.

Appendix II

II.1 FASTENERS, SEALS AND GASKETS

Fluid Hydraulic and the implements for which they provide power are held together by fasteners, seals and gaskets systems.

There are many kinds of fasteners and seals kits for hydraulic systems. Some are common and others are designed to perform special functions. During service fasteners may be exposed to conditions such as heating and cooling, cyclic loading, tensile and shearing loads, shock loads, corrosion and vibration.

It is common to see the various types of fasteners:

1. Screws

2. Set Screws

3. Bolts, Washers (or Spring Washers) & Nuts

4. Lock Nuts

This group of fasteners that hold parts together by passing through one part and threading into another part. It is very common to notice that many of these fasteners used for hydraulic services are of hexagonal heads and or Cap heads (hexagonal). Bolts are graded according to tensile strength and are noticeable by examining the bolts markings on the surface of the hexagonal face.

For SAE standards, grade 5 bolt has three marks on the head and grade 6 has four marks. Metric standard types are denoted by a number followed by an alphabet, eg. 5D or 10K.

Flat washers are used to provide a wider bearing surface for a bolt head or nut. When tightening a bolt against a softer material, such as Aluminium or brass or copper, the head may gradually become imbedded in the surface. This may cause the fastener to become loose during the use, especially where vibration environment. Sometimes, lock washers are also incorporated to prevent loosening of bolts. There are many others depend on it purposes.

Pins

Pins are used to either retain parts in a fixed position or to preserve alignment of parts. There are several types of pins that may be found on some hydraulic valves or pump assemblies.

Cotter Pins

Cotter pins are sized by a nominal dimension, such 3/32". The size of the hole for cotter pin should be slightly larger than the nominal size of the pin. Cotter pins are used to lock castle nuts and secure clevis pins (commonly found on cylinders' clevis pin). There are many standard sizes and lengths which most often must be trimmed to desired length.

Clevis pins

Clevis pins function as an axil so a part can swivel on it. It requires a flat washer and cotter pin to prevent the part from sliding off the pin.

Dowel Pins

Dowel pins are used for alignment and usually fit very snugly. They are heat treated and hardened. The dowel pin is pressed into a hole with an interference fit. This means the mating part has a matching hole that fits closely to the pin, but allows the part to be assembled or disassembled easily. They are usually found in the control valves as seat locator pin, multiple casing pump blocks and many others hydraulic components and assemblies' parts where orientation are critical.

Grooved Pins

Grooved pins are driven into an interference hole. The groove cuts into the wall of the hole and secures the pin. There are seven types of grooved pins. Each type is shaped differently and each has a different size and shape of groove to serve its' purposes.

Spring Tension Pins

Retaining rings are quite similar to grooved pins, except that it is not a solid pin. It is formed by spring steel and has a hollow center. The sectional view is like a circular 'C' and usually also driven into an interference hole. The spring tension holds the joint of two parts in place without slipping out. These types of pins are found commonly near the pivot of the manual lever of manual directional control valves.

Retaining Rings

Retaining rings are circular spring steel that fit externally or internally into a groove of a part. For example, they may be placed in a groove that is machined into the surface of a shaft, or internally in a groove in a cylindrical hole.

The purpose of an external retaining ring is to prevent movement of a shaft beyond a point through a hole such as in a bearing. At the same time the retaining ring does not prevent the shaft from rotating. Internal retaining rings prevent a shaft from travelling beyond the retaining ring located in its groove in the cylindrical part. The internal retaining rings are commonly used in the welded type hydraulic cylinders. It is usual to locate one of this at the cap head of the cylinder.

Most retaining rings must be installed and removed with a retaining ring tool.

Sealants, seals, gaskets and o-rings

Sealants, seals, gaskets and o-rings are compulsories item for hydraulic components and systems. Without these, high pressure systems are impossible.

Sealing Materials is an art

So selecting the right material is an art because most applications are compromises, such as resilience versus stiffness, softness versus hardness, low temperature versus high temperature and performance versus cost.

Nitrile Rubber Seals

Nitrile (also known as Buna N) seals are widely used in industry because of excellent resistance to oil, a good practical range of working temperatures (-40°F to as high as 275°F), and low cost. Oil on the surface of nitrile rubber even helps prevent oxidation and aging, which is a problem when nitriles are exposed to air and sunlight.

Performance in fluid power system is generally excellent. For example, Parker Seal Company, ran a series of brutal tests to determine how much fluid pressure a common nitrile O-ring can withstand without extruding into the gap. To make the tests more meaningful, they held the temperature at 160°F and cycled the pressure 100,000 times at each value.

The hardness for elastomers is measured by Shore durometer. The most typical nitrile rubber seals have hardness range of 70, 80 and 90.

Urethane Resists Abrasion

Polyurethane elastomers, commonly called urethanes, bridge the gap between synthetic rubbers and plastics. They have a unique combination of hardness, resilience, and load-bearing capacity plus exceptional abrasion resistance. The temperature rating is moderate: 200°F. The mixes are proprietary and result in a wide range of available properties. The material can be formulated for extrusion, moulding, casting or machining. Hardness can be as soft as a pencil eraser or as hard as bowling ball.

Urethane can significantly outwear most rubbers and plastics and even some metals. Also, the coefficient of friction is fairly low and gets even lower as hardness is increased. Urethane seals enjoy wide application as rod scrapers and other dynamic seals in harsh environments. Some grades of urethane are used in high-pressure hydraulic hose, which further attests to its strength and its ability to take abuse

Viton Fluoroelastomer

Dupont supplies this particular material to the seal manufacturers. It's fairly expensive but is being recognized for certain performance advantages that make it economical in numerous applications. It even replaces low-cost seals where reliability outweighs initial cost.

The normal maximum operating temperature is 400 to 600°F for intermittent service, and some hardness and resilience are retained at that high level. Even if the high operating temperature isn't needed, the hot tear strength of Viton makes it easy to strip from a mould without damage, enabling intricate shapes to be formed. Dynamic properties and low-temperature performance are also good.

Teflon Flouroplastic

Teflon (DuPont) and Halon and some similar flouroplastics from other companies have two key properties that make them interesting for sealing. They work at temperatures from cryogenic to 600°F, and they are self-lubricating. The prime disadvantages are creep (cold flow) and limited techniques for forming the parts. The creep is taken care of by designing the seal to either trap the plastic flow and thus contain it at the seal area or to add a metal or elastomer spring to maintain sealing force despite the creep. Teflon seal blending and manufacture is an art similar to powder metallurgy.

Dual-Material Seals

Grover/Universal Seal solves conflicting requirements of resilience on the one hand and hardness on the other with uni-ring, a dual-durometer one-piece sealing ring that does both. The sealing surface of the seal is hard (97 durometer) urethane; the backup is softer (45 durometer) urethane.

There are also many other forms of combination, for example inserts a nitrile rubber O-ring within a urethane U-cup to achieve a spring-loaded rod and piston seal. The O-ring assures sealing at low fluid pressure, and the sealing force increases as pressure rises.

O-rings

O-rings are another wonder piece of man-made art, humble looking and yet do wonder for many things.

An O-ring is a one-piece moulded elastomeric seal with circular cross-section that seals by distortion of its resilient elastic compound. Dimensions of O-rings are given in ANSI/SAE AS568A., Aerospace size standard for O-rings. Its have both imperial inches and metric mm sizes. The standard ring sizes have been assigned identifying dash number that in conjunction with the compound (ring material), completely specifies the ring. Although the ring sizes are standardized, however ANSI/SAE AS568A does not cover the compounds used in making the rings. In this case, manufacturers will use different designations to identify various ring compounds. For example, 230-8307 represents a standard O-ring of size 230 (2.484" ID by 0.139" width, or thickness of O-ring) made with compound 8307, a general-purpose nitrile compound.

O-Rings' Materials

Thousands of O-ring compounds have been formulated for specific applications. Some of the most common types of compounds and their typical applications are given in Table II.1. The Shore A durometer is the standard instrument used for measuring the hardness of elastomeric compounds. The softest O-rings are 50 and 60 Shore A and stretch more easily, exhibit lower breakout friction, seal better on rough surfaces, and need less clamping pressure than harder rings. For a given squeeze, the higher the durometer hardness of a ring, the greater is the associated friction. This is because a greater compressive force is exerted by hard rings than soft rings.

O-rings Installations

When properly installed in a groove, an O-ring is normally slightly deformed so that the naturally round cross-section is squeezed diametrically out of round prior to the application of pressure. This compression ensures that under static conditions, the ring is in contact with the inner and outer walls enclosing it, with the resiliency of the rubber providing a zero-pressure seal.

When pressure is applied, it tends to force the O-ring across the groove, causing the ring to further deform and flow up to the fluid passage and seal it against leakage, as in Figure. II.1 As additional pressure is applied, the O-ring deforms into a 'D' shape, as in Figure. II.2. If the clearance gap between the sealing surface and the groove corners is too large or if the pressure exceeds the deformation limits of the O-ring material (compound), the O-ring will extrude into the clearance gap, reducing the effective life of the seal.

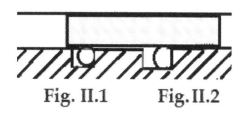

Fig. II.1   Fig. II.2

For very low-pressure static applications, the effectiveness of the seal can be improved by using a softer durometer compound or by increasing the initial squeeze on the ring, but at higher pressures, the additional squeeze may reduce the ring's dynamic sealing ability, increase friction, and shorten ring life.

There are many factors to consider for successful sealing conditions under certain specific applications. The initial diametric squeeze of the ring is very important in the success of an O-ring application. The squeeze is the difference between the ring width W and the gland depth and has a great effect on the sealing ability and life of an O-ring application. The ideal squeeze varies according to the ring cross-section; with the average being about 20% that is the ring's cross-section W is about 20 per cent greater than the gland depth F (groove depth plus clearance gap). The groove width is normally about 1.5 times larger than the ring width W. When installed, an O-ring compresses slightly and distorts into the free space within the groove, slightly elliptical in shape.

Table II.1 - O-rings material

S/N	O-ring Compounds	Descriptions
1.	Nitrile	General-purpose compound for use with most petroleum oils, greases, gasoline, alcohols and glycols. Effective temperature range is about -40°F to 250°F. Excellent compression set, tear and abrasion resistance, but poor resistance to ozone, sunlight, and weather. Higher-temperature nitrile compounds with similar properties are also available.
2.	Hydrogenated Nitrile	Similar to general-purpose nitrile compounds with improved high-temperature performance, resistance to aging and petroleum product compatibility.
3.	Polyurethane	Toughest of the elastomers used for O-rings, characterized by high tensile strength, excellent abrasion resistance, and tear strength. Compression set and heat resistance are inferior to nitrile. Suitable for hydraulic applications that anticipate abrasive contaminations and shock loads. Temperature service range of -65°F to 212°F.
4.	Fluorosilicone	Wide temperature range (-80°F to 450°F) for continuous duty and excellent resistance to petroleum oils and fuels. Recommended for static applications only, due to limited strength and low abrasion resistance.
5.	Fluorocarbon (Viton)	General-purpose compound suitable for applications requiring resistance to aromatic or halogenated solvents or to high temperature (-20° to 500°F with limited service to 600°F). Outstanding resistance to blended aromatic fuels, straight aromatics, halogenated hydrocarbon and other petroleum products..

II.2 Three Phase Current power supply system

3- phase Induction motors

Large hydraulic pumps are often driven by powerful and huge electric motors known as the three-phase induction motors. To drive this type of motors, the supply power must be of 3 phase system delivering 3 phase current with the respective voltage rated as KVA.

In AC motors, the input volt-ampere rating is derived from the current and voltage rating.

1. For Single phase motors, rated voltage = (rated volts)(rated amperes).

2. For Three phase motors, rated voltage = $\sqrt{3}$ (rated volts)(rated amperes).

Single phase motors are not self starting unless it was fitted with an auxiliary winding or couple with a starter (usually capacitors). Unlike single phase motors, three phase motors have 3 separate coils that are wound on to the soft iron stator core. Each coil is connected to a different phase denoted by R (red), Y (yellow) and B (blue). These are the identifying colours of the leads for the three phases. The 'R-Y-B' sequence is universally adopted to denote the emf in the yellow phase lags that in the red phase by a third of a cycle, and the blue phase lags that in the yellow phase by another third of a cycle.

ABOUT THE AUTHOR

M. Winston, BSc (Hons) from University of Glasgow, is an engineering consultant and designer for many years offering advanced processes and controls solutions for many industries including ship buildings, manufacturing services, oil and gas industries and pharmaceutical plants. He had helped many other diverse fields using non-traditional methods automating their plants' processes and facilities.